Innovative UV Technologies to Oxidize Organic and Organoleptic Chemicals

Awwa Research Foundation

About the Awwa Research Foundation

The Awwa Research Foundation (AwwaRF) is a member-supported, international, nonprofit organization that sponsors research to enable water utilities, public health agencies, and other professionals to provide safe and affordable drinking water to consumers.

The Foundation's mission is to advance the science of water to improve the quality of life. To achieve this mission, the Foundation sponsors studies on all aspects of drinking water, including supply and resources, treatment, monitoring and analysis, distribution, management, and health effects. Funding for research is provided primarily by subscription payments from approximately 1,000 utilities, consulting firms, and manufacturers in North America and abroad. Additional funding comes from collaborative partnerships with other national and international organizations, allowing for resources to be leveraged, expertise to be shared, and broad-based knowledge to be developed and disseminated. Government funding serves as a third source of research dollars.

From its headquarters in Denver, Colorado, the Foundation's staff directs and supports the efforts of more than 800 volunteers who serve on the board of trustees and various committees. These volunteers represent many facets of the water industry, and contribute their expertise to select and monitor research studies that benefit the entire drinking water community.

The results of research are disseminated through a number of channels, including reports, the Web site, conferences, and periodicals.

For subscribers, the Foundation serves as a cooperative program in which water suppliers unite to pool their resources. By applying Foundation research findings, these water suppliers can save substantial costs and stay on the leading edge of drinking water science and technology. Since its inception, AwwaRF has supplied the water community with more than $300 million in applied research.

More information about the Foundation and how to become a subscriber is available on the Web at **www.awwarf.org**.

Innovative UV Technologies to Oxidize Organic and Organoleptic Chemicals

Prepared by:
Karl G. Linden and **Charles M. Sharpless**
Department of Civil and Environmental Engineering, Duke University, Durham, NC 27708-0287

Susan A. Andrews
Department of Civil Engineering, University of Waterloo, Waterloo, Ontario, Canada

Khalil Z. Atasi and **Vinod Korategere**
Tetra Tech, Inc., Detroit, MI

Mihaela Stefan
Trojan Technologies Inc., London, Ontario, Canada
and

I. H. Mel Suffet
Environmental Science and Engineering Program, School of Public Health, UCLA, Los Angeles, CA

Jointly sponsored by:
Awwa Research Foundation
6666 West Quincy Avenue, Denver, CO 80235-3098
and

U.S. Environmental Protection Agency
Washington D.C.

Published by:

Awwa Research Foundation
Advancing the Science of Water®

IWA Publishing

DISCLAIMER

Published by IWA Publishing, Alliance House, 12 Caxton Street, London SW1H 0QS, UK

Telephone: +44 (0) 20 7654 5500; Fax: +44 (0) 20 7654 5555; Email: publications@iwap.co.uk;
Web: **www.iwapublishing.com**

AwwaRF report number 91033F. Originally published by AwwaRF for its subscribers in 2004.
IWA Publishing version published 2005.

ISBN 1 84339 919 9

CONTENTS

LIST OF TABLES

LIST OF FIGURES

FOREWORD

The Awwa Research Foundation is a nonprofit corporation that is dedicated to the implementation of a research effort to help utilities respond to regulatory requirements and traditional high-priority concerns of the industry. The research agenda is developed through a process of consultation with subscribers and drinking water professionals. Under the umbrella of a Strategic Research Plan, the Research Advisory Council prioritizes the suggested projects based upon current and future needs, applicability, and past work; the recommendations are forwarded to the Board of Trustees for final selection. The foundation also sponsors research projects through the unsolicited proposal process; and various joint research efforts with organizations such as the U.S. Environmental Protection Agency, the U.S. Bureau of Reclamation, and the Association of California Water Agencies.

This publication is a result of one of these sponsored studies, and it is hoped that its findings will be applied in communities throughout the world. The following report serves not only as a means of communicating the results of the water industry's centralized research program but also as a tool to enlist the further support of the nonmember utilities and individuals.

Projects are managed closely from their inception to their final report by the foundation's staff and large cadre of volunteers who willingly contribute their time and expertise. The foundation serves a planning and management function and awards contracts to other institutions such as water utilities, universities, and engineering firms. The funding for this research effort comes primarily from the Subscription Program, through which water utilities subscribe to the research program and make an annual payment proportionate to the volume of water they deliver and consultants and manufacturers subscribe based on their annual billings. The program offers a cost-effective and fair method for funding research in the public interest.

A broad spectrum of water supply issues is addressed by the foundation's research agenda: resources, treatment and operations, distribution and storage, water quality and analysis, toxicology, economics, and management. The ultimate purpose of the coordinated effort is to assist water suppliers to provide the highest possible quality of water economically and reliably. The true benefits are realized when the results are implemented at the utility level. The foundation's trustees are pleased to offer this publication as a contribution toward that end.

The research presented in this report can be viewed as a complement to the recent AwwaRF sponsored research in UV disinfection. Here, degradation of organic and organoleptic contaminants using UV based photolysis and UV/hydrogen peroxide advanced oxidation is investigated. Bench scale studies were performed to develop fundamental information on contaminant degradation, UV byproduct formation, and evaluate the impact of water quality on these outcomes. Pilot scale studies were also run to develop information of practical importance. Although the use of UV for contaminant degradation will not significantly occur during disinfection, this research has demonstrated that UV and UV based advanced oxidation can be powerful treatment processes for many chemicals of concern to the drinking water industry, without the harm of producing byproducts. Thus, those utilities installing UV for disinfection could expand these systems to achieve multiple water treatment objectives (i.e., contaminant degradation) through use of higher UV doses.

Walter J. Bishop
Chair, Board of Trustees
Awwa Research Foundation

James F. Manwaring, P.E.
Executive Director
Awwa Research Foundation

ACKNOWLEDGMENTS

The authors of this report are indebted to the following water utilities, companies, and individuals for their cooperation and participation in this project:

- City of Detroit, Michigan. Detroit Water and Sewerage Department, including Pamela Turner, T. Shajahan, Pawan Kapila, and Southwest WTP staff.
- Mannheim Water Treatment Plant, Kitchener ON Canada, Franklyn Smith.
- City of Durham, NC. Brown water Treatment Plant, Terry Rolan.
- Metropolitan Water District of Southern California, La Verne, CA, Mr. Alexander A. Mofidi.
- Trojan Technologies Inc, London ON Canada, Linda Gowman.
- Calgon Carbon Corporation, Pittsburgh, PA, Wayne Lem and Keith Bircher.
- Hanovia, Slough, U.K., Jon McClean.
- Pulsar Environmental Systems, Auburn, CA, Jim Bender.
- Jim Bolton, Bolton Photosciences Inc, Edmonton AB Canada.

Research carried out at Duke University could not have been possible without the help of students and laboratory staff. Dr. Charles Sharpless, a post-doctoral research fellow at the time of this research, was in charge of development of analytical methods, dose measurement for all bench reactors, and carrying out the experimental protocols for the work done at Duke. He was assisted by Ph.D. candidate Changlong Wu and Laboratory Manager Ms. Dwina Martin.

Dr. Wenjun Liu carried out UV irradiations, performed and supervised UFC testing and DBP analysis, and studied means to quench residual hydrogen peroxide concentrations prior to subsequent testing during his tenure as a post-doctoral fellow the University of Waterloo. Erin Moffat, project chemist, also performed much of the bench-scale DBP testing and was instrumental in keeping the pilot-scale UV reactor operational. Also acknowledged are the efforts of several undergraduate research assistants.

Work was completed at the University of California, Los Angeles by Dr. Mel Suffet and graduate students Jose Maude (Department of Statistics) and Suzan Givens, Navreet Aujla and Janel Grebel, Department of Environmental Health Sciences School of Public Health.

Work completed at Tetra Tech MPS was assisted by Theping Chen and Mohamed Siddiqui, Ph.D., P.E.

We gratefully acknowledge the technical and financial assistance of Trojan Technologies Inc., London, Ontario, Canada, throughout both the research and pilot phases of the project. Special thanks to Dr. Linda Gowman, Vice-president Research and Development and Dr. Ted Mao, Core Sciences Manager, for their kind and prompt support, and Mr. Alan Royce, P.Eng., M.Sc., for his contribution on the pilot testing of the UV/H_2O_2 AOP.

Direction provided by the Project Advisory Committee—including Sam Hayes, U.S. EPA Cincinnati, OH; Jim Malley, University of New Hampshire, Durham, N.H.; Gil Crozes, Carollo Engineers, Boise, ID; and Dixie Fanning, Charleston Commissioners of Public Works, Charlestown, SC — and AwwaRF project officers Albert Ilges and Kenan Ozekin is greatly appreciated.

EXECUTIVE SUMMARY

INTRODUCTION

Ultraviolet (UV) irradiation for primary disinfection of surface water supplies is rapidly gaining interest as a water treatment practice in the United States (U.S.). Recent regulations are now requiring lower chlorinated disinfection by-products (DBPs) concentrations in finished waters and inactivation of chlorine resistant pathogens such as *Cryptosporidium parvum.* Due in part to studies indicating that UV is effective for the inactivation of *Cryptosporidium parvum*, the U.S. Environmental Protection Agency (U.S. EPA) is now recommending UV as a compliance technology for primary disinfection of surface waters in the Stage 2 Disinfectants and Disinfection Byproducts Rule (D/DBPR) and Long Term 2 Enhanced Surface Water Treatment Rule (LT2ESWTR). One area of research that needs to be explored for application of UV to disinfection of surface waters is the potential oxidative impacts of UV light, especially high irradiance polychromatic UV sources, on organic and organoleptic compounds present in source waters. Information on the potential for UV technologies, alone and as advanced oxidation processes (AOPs) in combination with hydrogen peroxide, to either oxidize organic compounds found in some source waters such as pesticides, herbicides, volatile organic compounds (VOC), and taste and odor compounds and/or produce DBPs is needed to allow utilities to make informed decisions on how best to apply UV technology for potable water.

Many small and large utilities are already in the stages of developing and applying UV technologies to their treatment processes as a control technology for *Cryptosporidium parvum* and other pathogens. However, not much is known about the potential formation of UV DBPs nor the oxidative capabilities of UV and AOPs for use in drinking water. Questions such as "What types of changes in organic compounds are expected with UV treatment?" and "How will taste and odor be impacted by UV treatment?" have not yet been fully addressed on a bench scale nor significantly tested at the pilot or full-scale where performance and operational costs can be estimated. Now that UV is going to be used as a compliance technology to meet upcoming federal rules, these types of questions need to be addressed so that utilities can make informed decisions on their long-term needs for disinfection, control of DBPs, and possible implementation of UV for treatment of other water pollutants.

Utilities will also need to choose among a number of different UV technologies including low lressure (LP), low pressure high output (LPHO), medium pressure (MP), and pulsed-UV (P-UV) applications. The comparison of UV technologies and the ability of each to degrade various pollutants as described in this research report will aid in the comparative analyses utilities will need to perform.

PROJECT APPROACH

Given the current understanding of the microbial inactivation capabilities of UV, the purpose of the proposed research was to further investigate the chemical risks (and potential benefits) associated with implementation of emerging UV technologies. The overall goal of the research was to examine the effect of the application of UV technologies (with fundamental differences in output spectrum, high and low irradiance, and continuum of energy output) and UV advanced oxidation applications on specific water pollutants such as taste and odor causing chemicals, levels of contaminants (e.g., pesticides/herbicides/volatile organic compounds

(VOC)) and DBP formation (trihalomethones - THMs, haloacetic acids - HAAs, aldehydes, carboxylic acids).

The research was first geared toward evaluating the efficacy of different UV technologies including low pressure mercury (Hg) lamps (LP), medium pressure Hg lamps (MP) and xenon based pulsed UV lamps (P-UV) for treatment of different classes of contaminants in waters of different qualities as well as formation of DBPs both from the UV technologies themselves and from post-chlorination practices. The UV technologies were evaluated alone and with the addition of hydrogen peroxide (H_2O_2) for evaluation of advanced oxidation process conditions. The relative rates of contaminant destruction using UV and the UV AOP were determined and measures of the electrical energy needed for contaminant destruction were made. Finally, LP and MP UV technologies were pilot-tested at two different utility field sites for selected contaminant treatment to develop operational experience and preliminary cost data.

KEY FINDINGS

The investigation of the UV photolysis and oxidation of the organic and organoleptic contaminants tested resulted in a number of key findings of importance to the water treatment industry. Based on our research, it is quite clear that at typical UV doses used for disinfection of drinking water (approximately 40 mJ/cm^2), there will be almost no apparent chemical degradation benefits for pollutants of concern. In fact, for the contaminants researched in this study, to achieve one log (90%) of degradation via UV photolysis, the UV dose would have to be at least 500 mJ/cm^2 (more than ten times the "typical" disinfection dose of 40 mJ/cm^2). For some of the contaminants, even doses as high as 2000 mJ/cm^2 were not effective for this level of degradation. Thus, the effectiveness of UV for contaminant degradation is also contaminant-specific. For some contaminants, such as NDMA, designing a UV system to achieve both disinfection and contaminant destruction would appear feasible. If an oxidant such as hydrogen peroxide (H_2O_2) is added to the water in the presence of UV light, then the UV reactivity of many compounds greatly increases as shown for atrazine in Figure ES.1.

There was no major impact of water quality on the UV-based degradation processes tested once all the proper corrections for UV absorbance were made in the delivered dose measurement. Thus, the effect of such factors as TOC, which can alter the UV absorbance, was not counted in these effects but appeared to be of importance on the bench reactor scale and the pilot reactor scale where the UV output was fixed and water absorbance changes affected the UV transmission. Both pH and temperature played minor roles in affecting degradation. The addition of H_2O_2 played a more significant role for some contaminants (e.g. MIB) than for others (e.g., NDMA). On the pilot scale, the addition of H_2O_2 also played a major enhancing role for treatment of taste and odor compounds.

Three different UV sources were tested: LP, MP and P-UV lamps. On a per dose basis, the use of a MP UV source often improved treatment for most of the pollutants tested. However, on the basis of electrical efficiency, the LP UV source was often the most efficient for degradation. Two P-UV sources were also tested. Allowing for experimental uncertainties in the UV dose measurement for the pulsed lamps, they appeared to perform similar to or slightly better than the MP lamps for both atrazine and NDMA contaminants on a UV dose basis. Although the P-UV lamp results were intriguing, comparison with the mercury vapor lamps was complicated by several factors, including variable energy efficiency of different pulsed lamps and difficulties in accurately quantifying the UV dose for these lamps. Further testing of the pulsed UV sources is warranted. These systems may prove to be viable alternatives in the near future.

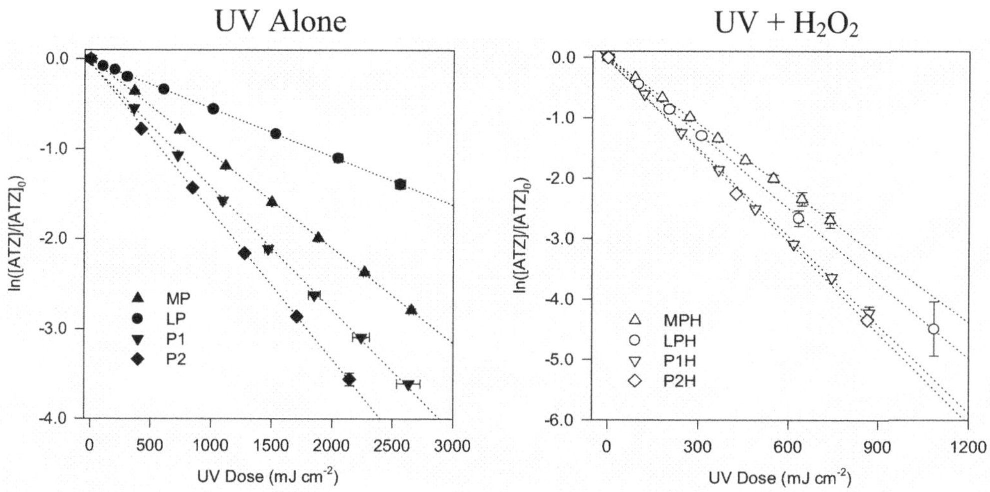

Figure ES.1 Kinetic plots for atrazine UV photolysis with and without H_2O_2 addition with LP, MP and P-UV lamps

Addition of H_2O_2 leads to a marked increase in removal rates for atrazine with all the lamps, and adding oxidant decreased the electrical requirements for atrazine removal. With NDMA, H_2O_2 either had no effect or actually decreased the electrical efficiency of the process due to screening of the directly effective UV light. The treatment of MIB and geosmin was virtually non-existent without addition of H_2O_2. Thus, use of this additive is essential for UV treatment effectiveness against these contaminants. UV degradation of metolachlor, atrazine, diazinon, carbaryl, and toluene were all tested with and without H_2O_2 but these contaminants degraded much more rapidly in the presence of peroxide. In general, the addition of H_2O_2 in a UV pollutant treatment process greatly enhances the degradation of many contaminants.

The potential for UV, used for disinfection and for pollutant degradation, to form direct disinfection byproducts or affect the formation of chlorinated byproducts following post chlorination is essential information for ultimate acceptance of UV technologies. The formation of aldehydes and carboxylic acids as direct byproducts and trihalomethanes and haloacetic acids as post chlorination byproducts was tested. These products were monitored following UV treatment by LP, MP and P-UV sources, in the presence and absence of H_2O_2 addition. UV treatment at doses up to 140 mJ/cm^2 caused no measurable changes in any of the byproducts for any of the lamp types as illustrated in Figure ES.2 for THMs. Furthermore, at doses used for pollutant degradation (~1000-2000 mJ/cm^2), there were no changes in any of the disinfection byproducts measured that would be a concern for public health. This was true both in the presence and absence of H_2O_2. Thus, there is no apparent concern for disinfection byproduct formation from UV treatment processes.

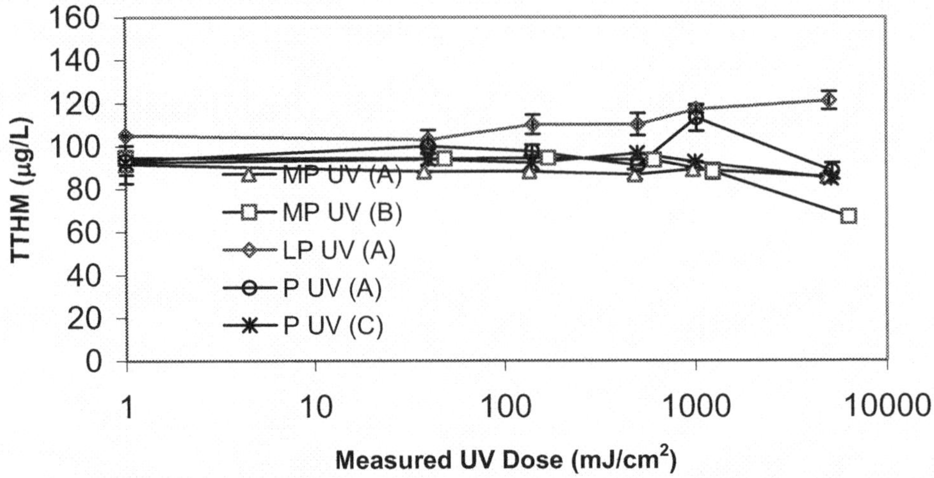

Figure ES.2 TTHM concentrations from UV irradiation of synthetic water followed by UFC testing (averages and standard deviations)

Among the suite of contaminants tested in the bench scale reactor, NDMA was by far the fastest to degrade under UV alone. Metolachlor was then the most reactive, followed by atrazine and diazinon, all of which had appreciable rates of direct photolysis. Carbaryl and toluene were both very unreactive with UV alone, as were MIB and geosmin. With the addition of H_2O_2, all of these contaminants (except for NDMA in some cases) decayed much more rapidly. Interestingly, the costs associated with treatment of the taste and odor compounds IPMP, geosmin, and MIB with UV/H_2O_2 are very comparable to those for treating NDMA, both with LP and MP lamps as summarized in Table ES.1. Of the remaining compounds, metolachlor is the most susceptible to UV/H_2O_2 treatment, and is more efficiently treated by LP/H_2O_2 than MP/H_2O_2. However, treating this compound will require approximately two to three times more electricity than NDMA for a given level of removal. Much higher costs can be expected for treating atrazine (ATZ) and diazinon, and higher costs still for treating toluene and carbaryl.

Table ES.1
Cost in $U.S. to treat (1-log removal) 1,000 gal of Durham water with UV/H_2O_2 (5 mg/L)

	ATZ	Carbaryl	Diazinon	Metolachlor	NDMA	Toluene	IPMP	Geosmin	MIB
MP	0.39	0.58	0.47	0.25	0.11	0.57	0.09	0.07	0.13
LP	0.44	0.62	0.36	0.30	0.08	0.64	0.09	0.05	0.14

Regarding the use of UV light for contaminant degradation, this research has provided some fundamental bases for selection of different UV and UV/oxidation technologies given the different types of contaminants faced by the water industry. Clearly, there is no "one size fits all" answer with UV. For instance, comparing the treatment of NDMA, atrazine and MIB, an engineer may come up with different ideal UV-based solutions for each contaminant type because each has a unique absorbance spectrum and quantum yield (relating to the potential for photolysis with different lamp types) as well as a unique hydroxyl radical reaction rate constant (relating to the effectiveness of UV/peroxide oxidation).

The side-by-side comparison of UV technologies on the bench- and pilot-scale for the ability of each to degrade compounds will aid designers in the types of analyses utilities will need to perform when deciding upon which technology to use. The comparative treatment efficiencies for different contaminants tested in the study help to recognize where some of the strengths and weaknesses of UV technology are with relation to certain pollutants. Thus, the knowledge generated from this research can be used by both regulators and utilities alike in deciding how best to approach issues surrounding disinfection byproducts in drinking water and removal of unwanted organic and organoleptic compounds from water.

CHAPTER 1
INTRODUCTION

BACKGROUND

Ultraviolet (UV) irradiation for primary disinfection of surface or ground water supplies is not presently a widely utilized water treatment practice in the United States (U.S.). Recent regulations are now requiring lower chlorinated disinfection by-products (DBPs) and inactivation of chlorine resistant pathogens such as *Cryptosporidium parvum.* Due in part to studies indicating that UV is effective for the inactivation of *C. parvum*, the U.S. Environmental Protection Agency (U.S. EPA) is now including UV as a compliance technology for primary disinfection of surface waters in the Stage 2 Disinfectants and Disinfection Byproducts Rule (D/DBPR) and Long Term 2 Enhanced Surface Water Treatment Rule (LT2ESWTR). One area of research that needs to be explored for application of UV to disinfection of surface waters is the potential oxidative impacts of UV, especially high irradiance polychromatic UV sources, on organic and organoleptic compounds present in source waters. Information on the potential for UV technologies, alone and as advanced oxidation processes (AOPs) in combination hydrogen peroxide, to either oxidize organic compounds found in some source waters such as pesticides, herbicides, volatile organic compounds (VOC), and taste and odor compounds and/or produce DBPs is needed to allow utilities to make informed decisions on how best to apply UV technology for potable water.

Significance to Water Utilities

Many water utilities are already in the stages of developing and applying UV technologies to their treatment processes as a control technology for *C. parvum* and other pathogens. However, not much is known about the potential formation of UV DBPs nor the oxidative capabilities of UV and AOPs for use in drinking water. Questions such as "Where is UV best applied in a treatment train?"; "What types of changes in organic compounds are expected with UV treatment?"; "How will taste and odor be impacted by UV treatment?", have not yet been fully addressed on a bench scale nor significantly tested at the pilot- or full-scale, where performance and operational costs can be estimated. Now that UV is going to be used as a compliance technology to meet upcoming federal rules, these types of questions need to be addressed so that utilities can make informed decisions on their long term needs for disinfection, control of DBPs, and possible implementation of UV for treatment of other water pollutants.

Utilities will also need to choose among a number of different UV technologies including low pressure (LP), low pressure high output (LPHO), medium pressure (MP), and pulsed-UV (P-UV) applications. The comparison of UV technologies and the ability of each to oxidize various compounds as described in this research report will aid in the comparative analyses utilities will need to perform.

UV Technologies Overview

Although there are three principal UV technologies available on the market for disinfection, to date, the majority of the published literature on UV disinfection of drinking water has focused on the use of continuous wave UV irradiation from LP mercury vapor lamps. Other

UV technologies include continuous wave LPHO and MP mercury vapor lamps, and P-UV xenon arc lamps. The characteristics of typical LP-, LPHO-, MP-, and P-UV lamps are presented in Table 1.1. Comparison of the spectral outputs in the germicidal range of the LP, MP, and P-UV sources used in this study are presented in Figures 3.2 and 3.3.

Table 1.1
Characteristics of typical LP, LP-HO, MP and P-UV lamps

Characteristic	Low pressure / LP-HO	Medium pressure	Pulsed-UV
Emission	monochromatic (85-90% at 253.7 nm)	polychromatic (185-600 nm)	polychromatic (185-800)
Mercury vapor pressure	0.1 to 10 Pa	50 to 300 kPa	non-mercury
Operating temperature (°C)	30 to 50 / 60 to 100	600 to 900	8,000 to 25,000 °K
Arc length (cm)	15 to 200 / 15 to 200	10 to 200	10 to 100
Lifetime (hours, thousands)	8 to 12 / 7 to 10	3 to 8	>10^8 pulses
Efficiency (200-300 nm, %)*	35 to 40 / 30 to 35	12-16%	<20% (varies)
Light irradiance (relative)	low / medium	high	high

*Ratio of (total germicidal UV power emitted by the Lamp)/(electrical power in from the wall at the plug)

The environmental advantages of disinfection with LP-UV lamps, such as the lack of formation of harmful DBPs (Malley et al. 1995), are becoming more apparent. However, there are several disadvantages that make them impractical for some treatment facilities. High lamp numbers and large system footprints are sometimes required in facilities with effluent volumes higher than 10 million gallons per day (mgd) (Cairns 1996, Kwan et al. 1996). High maintenance and labor costs may also be associated with cleaning large numbers of UV lamps with sleeve fouling. Thus, there may be interest in evaluating the use of higher irradiance UV sources such as LPHO-, MP- and P-UV lamps for larger scale drinking water applications.

Low-Pressure Ultraviolet Irradiation

LP- and LPHO-UV lamps characteristically emit monochromatic, low intensity radiation principally at 253.7 nm, within the germicidal wavelength range as defined by the DNA absorbance spectrum. The germicidal UV dose (*H*) for these lamps is typically calculated as the product of the volume averaged incident irradiance (*E*, mW/cm^2) at 253.7 nm and the time of exposure (*t*, seconds) resulting in units of millijoules per square centimeter (mJ/cm^2) for UV dose.

Medium-Pressure Ultraviolet Irradiation

MP-UV lamps are mercury-arc lamps that operate at up to a million times the pressure and at much higher temperatures than LP-UV lamps. The increased pressure causes emission of higher intensity radiation and a broadening of emission lines, so that the emission becomes quasi-continuous and polychromatic over the range from the far UV (185-nm) to the infrared (1,367-nm). Approximately 15 to 25 percent of the polychromatic radiation emitted is in the germicidal range (200-300 nm). Therefore, MP-UV lamps are much less germicidally efficient than LP-UV lamps. However, because the radiation emittance is so much greater, MP-UV lamps can emit approximately 60 to 600 times the germicidal irradiance of the LP-UV lamps. Two of the strongest emissions in this range are at 266 and 256 nm. This higher emittance translates into a comparatively smaller footprint for the MP-UV disinfection system (significantly fewer lamps for equivalent effect) and potentially lower operations and maintenance costs.

Pulsed-Ultraviolet Irradiation

P-UV lamps (xenon, argon, krypton or other inert gases or mixtures) are characterized by a high irradiance broadband emission spectrum within the 200-300 nm range. Their maximum wavelength and pulse rate can be adjusted electronically, making them of interest for various water treatment applications. P-UV lamps are essentially "blackbody" emitters, providing continuous emission spectra between the wavelengths of 200 and 1,000 nm. High-intensity, pulsed-UV radiation is produced by accumulating electrical energy in an energy storage capacitor over fractions of a second and releasing this stored energy by a high speed electrical switch to do work in a few microseconds (μs) (Dunn et al. 1995). The result is extremely intense flashes of polychromatic radiation that contain germicidally effective wavelengths. However, the actual germicidal irradiance emitted by these lamps is not currently well described. Recent interest has focused on the application to drinking water treatment. Prototypes of commercial systems are currently being researched for use in drinking water disinfection. However, because these systems are relatively new and most research has been proprietary, independent evaluations of system disinfection efficiency, the mechanisms behind disinfection, and an evaluation of DBP formation potential must be conducted.

Photochemical Processes

Direct Photolysis

Some pollutants undergo direct photolysis when irradiated with UV light. The free radicals, generated as a result of the photochemical dissociation, react in turn with dissolved oxygen, initiating oxidative degradation. A fundamental parameter for characterizing the efficiency of direct photolysis is the quantum yield, Φ, defined as the moles of product formed per mole of photons absorbed by the parent compound (a mole of photons is also called an einstein). The quantum yield thus represents the fraction of absorbed photons that cause chemical change. Direct photolysis is possible only if the absorption spectra of the pollutants overlap the emission spectrum of the UV lamp, and is efficient only if the quantum yield of the photochemical process is reasonably large. Most organic pollutants absorb strongly in the UV range, especially below 250 nm. In addition, the quantum yield of photodissociation tends to increase at lower wavelengths where the photon energy is larger. However, quantum yields for UV direct photolysis are generally small, and there are very few applications of this process in water treatment [an exception is the treatment of N-nitrosodimethylamine (NDMA)].

UV/H_2O_2 Indirect Photolysis

The application of UV to solutions containing H_2O_2 is very effective for oxidizing organic compounds. The basis for this method is that UV direct photolysis of H_2O_2 forms the strongly oxidizing hydroxyl radical ($^{\bullet}OH$). Hydrogen peroxide is a fairly weak absorber of UV light [ε_{max} (200 nm) = 198 M^{-1} cm^{-1}], but its photolysis occurs with a quantum yield of 1.00 (Baxendale and Wilson 1957). Hydroxyl radical reacts almost non-selectively with organic compounds either by H-abstraction or addition to double bonds, depending on the structure of the substrate. These reactions are characterized by the second-order rate constant, designated here as k_{OH}, and for many organic compounds k_{OH} is approximately 10^9 - 10^{10} M^{-1} s^{-1}, which indicates near diffusion control of the reaction rate. The resulting carbon-centered radicals react with dissolved oxygen at diffusion-limited rates, initiating rapid degradation to lower molecular weight compounds. Treatment with UV/H_2O_2 can lead to complete mineralization of organic pollutants. The efficiency of this process depends on the competition between H_2O_2 and pollutants both for the light and for $^{\bullet}OH$ radicals. Further competition for $^{\bullet}OH$ occurs with many natural scavengers, such as dissolved organic matter and bicarbonate (Peyton et al. 1998), making water quality an important factor to consider when comparing efficiencies.

LITERATURE REVIEW - CONTAMINANTS INVESTIGATED

Volatile and Semi-Volatile Contaminants (NDMA, toluene)

N-Nitrosodimethylamine (NDMA)

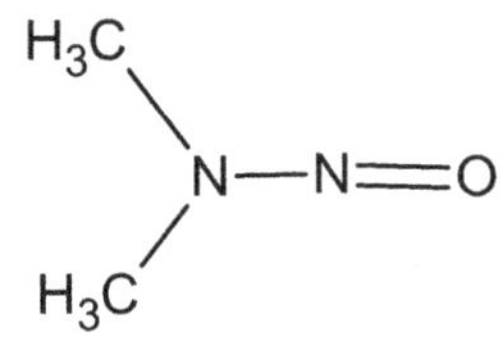

N-Nitrosodimethylamine (NDMA) is a semi-volatile organic compound that is extremely soluble in water. This, combined with its low Henry's Law constant [estimates vary between 3 x 10^{-7} and 3 x 10^{-5} atm-m^3/mol (Environment Canada 2001; ATSDR 1989)] makes it very persistent in water despite its volatility as a neat solvent. It has little industrial or commercial importance but occurs in the environment due to its inadvertent formation and release from such diverse industrial sources as tanneries, amine, pesticide and detergent manufacturing plants, rubber and tire manufacturers, fish and food processing plants, malt beverage producers, dye manufacturers, and cosmetics manufacturers. Another source of NDMA contamination is the aerospace industry, where it occurs due to decomposition of dimethylhydrazine in rocket fuel. In the environment, NDMA can form by the reaction of naturally occurring primary, secondary or tertiary amines with nitrites. Finally, recent reports indicate that NDMA can actually form during certain water treatment processes (Environment Canada 2001; Gerecke and Sedlak 2003).

Concentrations of NDMA in various waters range from low ng/L to tens of µg/L, and there is a growing concern with regard to the associated health effects of exposure. Hepatotoxicity is a primary systemic effect of NDMA, but it is also known animal carcinogen and is listed as a potential human carcinogen. The U.S. EPA does not currently regulate NDMA, but it is listed as a priority pollutant on the U.S. EPA National Priorities List. Furthermore, regulations and advisory levels are enforced in various states across the U.S. In Canada, the Ontario Ministry of the Environment has established an interim drinking water standard for NDMA of 0.009 µg/L (9 ppt).

As a chemical class, N-nitrosamines are thermally stable in aqueous solutions, resisting hydrolysis, biodegradation, and adsorption. Also, their high water solubility makes them unamenable to air-stripping, even though they tend to be volatile in their pure form. All N-nitrosamines are, however, photochemically labile. Therefore, UV and UV-based AOTs have been promoted for treating water contaminated with NDMA. In fact, the past decade has seen the development and installation of several full-scale UV reactors in North America for treatment of NDMA in drinking water, and ground and wastewaters.

While there have been numerous studies of NDMA photolysis in the past, most have been concerned with chemical mechanisms, and only a few have focused on aspects related to treatment in natural waters (Ho et al. 1996; Stefan et al. 1999; Geiger and Huber 1981; Kobylinski and Peterman 1979; Jobb et al. 1994; Cooper and Keller 1993). Also, no studies have previously been done to compare the effectiveness of different UV lamps and the effect of H_2O_2 on the process efficiency with different UV lamps.

Toluene

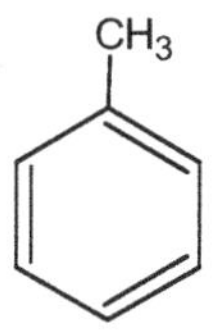

Toluene is a common organic solvent used in many industries including printing and painting and is a component of gasoline (approximately 6%) (ATSDR 2000). Although it is moderately soluble in water (530 mg/L (ATSDR 2000)), its high Henry's Law constant (6×10^{-3} atm m^3/mol, (ATSDR 2000)) indicates it will easily volatilize from water samples open to the atmosphere. Thus, it is not frequently detected at high levels in drinking water. However, it is often found at elevated concentrations (10 – 100 μg/L) near hazardous waste sites (ATSDR 2000). Presently, the U.S. EPA regulates toluene to a MCL of 1 mg/L.

Few data are available to assess sub-acute health effects of toluene following ingestion by humans. On the other hand, adverse health effects associated with inhalation of toluene are well documented and are primarily related to neurological damage and potential harm to the fetus in pregnant women (ATSDR 2000).

Treatment of toluene by UV and UV-based AOTs is not common, primarily due to the ease with which it can be air-stripped. Furthermore, it has a relatively low UV absorbance ($\varepsilon_{254} \cong 150\ M^{-1}\ cm^{-1}$). Thus, in this work it can be considered a difficult test-case for UV treatment. There are some reports of its removal from water by UV-based AOTs, many of which have examined by-products as well (Kang and Lee 1997; Mansour 1985; Sundstrom et al. 1989; Andrews et al. 1999), but these studies have all involved either monochromatic UV or wavelengths in the solar spectrum. Therefore, there is little information available concerning the effects of water quality and no information on the effects of different wavelengths below 200 nm for treating water contaminated with toluene.

Herbicides and Pesticides (atrazine, metolachlor, carbaryl, and diazinon)

Atrazine, 2-chloro-4-ethylamino-6-isopropylamino-s-triazine

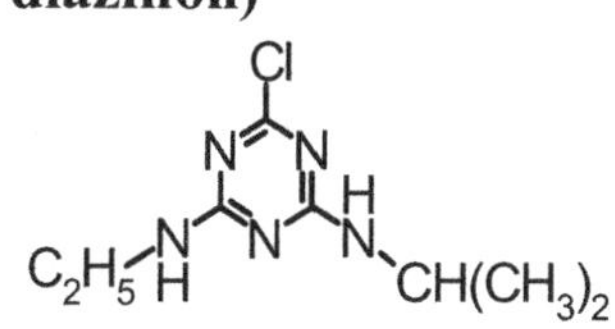

Atrazine (ATZ) is a widely used agricultural herbicide in many regions of the United States and Europe, and is commonly found throughout the entire hydrological cycle. The U.S. EPA currently regulates atrazine, along with other herbicides, to an MCL of 3 μg/L. The German regulations set a maximum legally permissible level of total herbicide concentration in drinking water at 0.5 μg/L and at 0.1 μg/L for any single herbicide (Nick and Schoeler 1995).

The risk of health effects to humans associated with exposure to ambient levels of ATZ in natural waters appears to be insignificant (AWWA 2003). However, a growing body of research literature indicates that these same levels of ATZ may have deleterious effects on aquatic wildlife. For example, fish and amphibians living in ATZ impacted waters in the U.S.

and U.K. display hermaphroditism and sex-reversal, suggesting that ATZ may be an endocrine disrupting compound (Hayes et al. 2002; Hayes et al. 2003).

Traditionally, ATZ removal from drinking water has been accomplished by adsorption on activated carbon. While effective, this treatment simply transfers the contaminant to a new phase, and disposal of the carbon then becomes an issue. Treatment by UV and UV/H_2O_2 offers the possibility of destroying the parent compound, when the issue then becomes considering the health effects of the resultant by-products. Bearing this in mind, a recent report has suggested that UV/H_2O_2 is a realistically viable option for removing ATZ and its by-products (Kruithoff et al. 2002). This is not so surprising when one considers the high reaction rate constant between ATZ and hydroxyl (·OH) radicals (2.6 x 109 M^{-1} s^{-1} (NDRL 2003)).

Numerous studies have been published on the photolysis of ATZ, and there exists a good understanding about the general mechanisms involved (e.g. Hessler et al. 1993; Rehberg et al. 1994; Nick and Schoeler 1995; Chan et al. 1992; Beltran et al. 1993 Beltran et al. 1996). However, the results of many of these studies are not easily understood in the context of UV treatment as they were either conducted in buffered laboratory water or involved sunlight-driven photolysis. When the studies did concern water treatment, the results were often presented in a very context-specific manner, making it difficult to generalize them to different systems and waters. Furthermore, no work has previously been published comparing different UV lamps in a manner that would allow prediction of process efficiency for new lamp types under different operating conditions (e.g., different water qualities, H_2O_2 levels).

Metolachlor, 2-chloro-N-(2-ethyl-6-methylphenyl)-N-(2-methoxy-1-methylethyl) acetamide

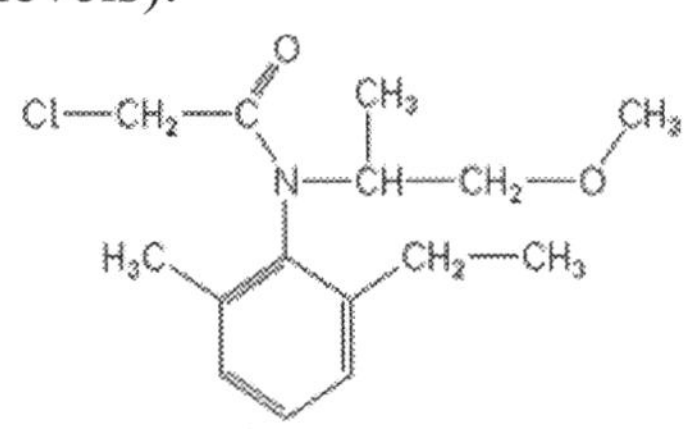

Metolachlor is one of the most widely used chloroacetanilide herbicides and occurs frequently in waters near agricultural areas (USGS 1999). It is moderately soluble in water (488 mg/L) (Agriculture Canada 1982). The U.S. EPA does not regulate metolachlor, but on the basis of limited animal studies that have indicated it to have carcinogenic potential (Ecobichon 1991) it is listed as a possible human carcinogen. It is also on the Candidate Contaminant List, meaning that EPA considers the collection of health, occurrence, and treatment data a priority so that its regulatory status can be properly reviewed. In the interim, EPA has set a lifetime health advisory for consumption in water at 0.1 mg/L (U.S. EPA 2002).

Volatilization, hydrolysis and photolysis are the major natural fate pathways for metolachlor in the environment. Although there are few reports on treating metolachlor with UV or UV-based AOTs, previous research does indicate that water quality can play a significant role, with photosensitization by natural organic matter, minerals, and nitrate all having been observed (Frimmel and Hessler 1994; Kochany and Maguire 1994; Lin et al. 1999; Mathew and Khan 1996; Wilson and Mabury 2000). It is also noteworthy that simulated solar irradiation (up to 4 J/cm^2) decreases the aquatic toxicity of metolachlor (Lin et al. 1999).

The U.S. EPA has not listed a preferred treatment method for metolachlor. It has a fairly high partition coefficient into soil (K_{OC} = 135, Kim and Feagley 1998) indicating that adsorption may be a useful procedure, and this is borne out by limited research (Griffini et al. 1999). However, treatment by UV also appears to be promising, based on the reported quantum yield of 0.34 for photolysis at 254 nm (Frimmel and Hessler 1994), which is comparable to that of

NDMA. No research on UV or UV/H_2O_2 treatment of metolachlor was located in the literature, but treatment by UV/O_3 and the Photofenton process have been studied (Kearney et al. 1987; Pignatello and Sun 1995). The results indicate that metolachor is susceptible to reaction with ·OH radicals, although no rate constant has been published for this reaction. Based on this, treatment by UV/H_2O_2 should also be effective.

Carbaryl (1-naphthyl N-methyl carbamate)

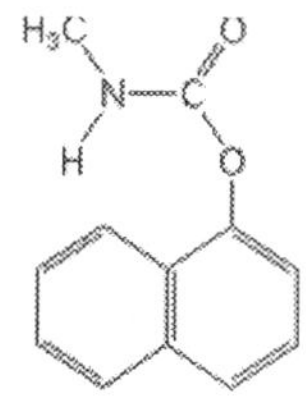

Carbaryl is a carbamate insecticide that is frequently detected in surface and groundwaters (USGS 1999). It is slightly soluble in water (40 – 50 mg/L) and has fairly large partition coefficients for both octanol (reported range of K_{OW} is 65 – 230) and organic carbon (reported range of K_{OC} is 200 – 450) (EXTOXNET). The U.S. EPA does not presently regulate carbaryl, but the agency has set a lifetime health advisory for consumption in water at 0.7 mg/L (U.S. EPA 2002). The major health effect associated with carbaryl is acute neurotoxicity due to its action as an acetylcholinesterase inhibitor (Ecobichon 1991).

The U.S. EPA has not listed a preferred treatment method for carbaryl, but there is some evidence that adsorption may be practicable (Hu et al. 1997). Photodegradation of carbaryl for water treatment has not previously been studied, but there is literature concerning its environmental photochemistry that indicates that carbaryl has a low photoreactivity (Wolfe et al. 1978; Yager and Yue 1988; Glod and Pawlak 1983). No reports exist that allow assessment of its photostability at wavelengths below 290 nm. There are reports, however, indicating that carbaryl may be susceptible to treatment by UV/H_2O_2. For example, Mabury and Crosby found that carbaryl was the most reactive pesticide towards ·OH radical out of ten pesticides tested (Mabury and Crosby 1996); its rate constant for reaction with •OH was estimated to be greater than 3 times that of atrazine.

Diazinon (O,O-diethyl-O-(2-isopropyl-6-methyl-4-pyrimidinyl) -phosphorothioate)

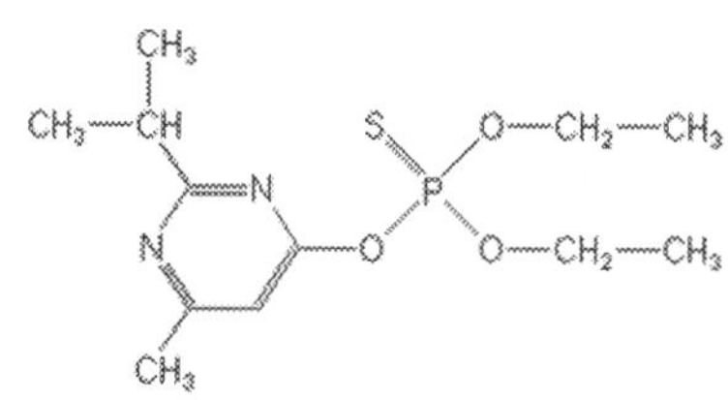

Diazinon is a commonly applied organophosphorus pesticide in both large-scale agriculture and home gardening. Due to its frequent use, it is one of the most commonly detected pesticides in surface and ground waters of the U.S. (USGS 1999). It is barely soluble in water (40 – 70 mg/L), has high partition coefficients for octanol ($10^{3.8}$) and organic carbon ($10^{2.3}$), and has a very low Henry's law constant (reports range from 3×10^{-9} to 10^{-6} atm m^3/mol) (ATSDR 1996). The primary adverse health effect associated with ingestion of diazinon is acute neurotoxicity; it is an acetylcholinesterase inhibitor, but unlike metolachlor it is irreversible. The U.S. EPA does not currently regulate diazinon, but it is on the Candidate Contaminant List. Some indication of its toxicity can be gleaned from the fact that EPA has set a very low lifetime health advisory for consumption of diazinon in water, 0.1 μg/L (U.S. EPA 2002).

There is presently no preferred treatment for water contaminated with diaznon, and there are very few studies examining treatment options. Because diazinon absorbs light within the 200-300 nm range it is possible that it may be degraded by UV-driven technologies. A few studies have examined this possibility, and it seems that UV and UV/H_2O_2 are both viable options (Doong and Chan 1998; Ku et al. 1998; Wan et al. 1994).

Taste and Odor Compounds

Geosmin, 2-methylisoborneol (MIB), and 2-isopropyl-3-methoxypyrazine(IPMP)

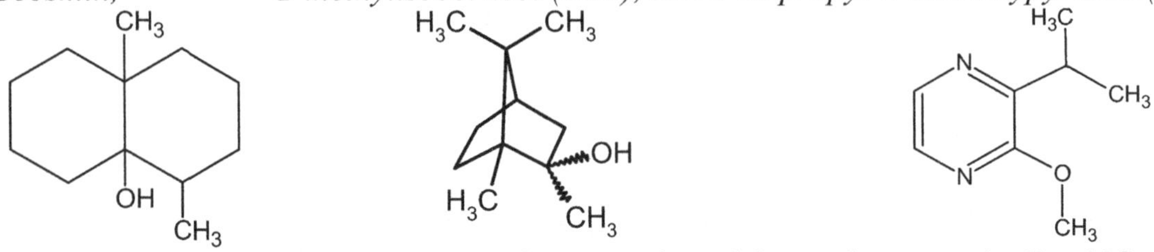

Three taste and odor compounds were selected for study: geosmin (1α-10β-dimethyl-9α-decalol), 2-methylisoborneol (MIB), and 2-isopropyl-3-methoxypyrazine (IPMP). Geosmin and MIB are often associated with odor incidents (Huck et al. 1995; Hattori 1988; Hrudey et al. 1995). Geosmin is described as having an earthy odor, whereas MIB's odor is described as musty (Persson 1988); these are produced by microorganisms, such as cyanobacteria in natural waters. The odor of IPMP is often described as that of decaying vegetation/or earthy. While these compounds are not considered harmful to human health, their presence has a negative impact on drinking water aesthetics. The threshold odor concentration for all three compounds is less than 10 ng/L (Morioka et al. 1993), making them problematic at extremely low concentrations.

Conventional treatment methods are fairly inefficient for the removal of geosmin and MIB, and advanced methods must generally be employed. Oxidants such as chlorine, potassium permanganate, and ozone provide some treatment (Hrudey et al. 1995; Lalezary et al. 1986). Ozone alone can remove 30-50 % of these compounds. However, enhanced removals are observed when AOTs are employed (Hrudey et al. 1995; Glaze et al. 1990; Lang et al. 1995; Andrews et al. 1995; Koch et al. 1992; Ferguson et al. 1990). When O_3, O_3/H_2O_2, or UV/O_3 are applied, a high pH is needed to promote ·OH formation. Of the three, IPMP appears to be removed most easily (AwwaRF and Lyonnaise des Eaux 1995).

STUDY SCOPE AND RESEARCH OBJECTIVES

There is great interest among water industries in North America and around the world in the use UV technology for inactivation of *Cryptosporidium parvum,* among other pathogenic microorganisms. These interests have important implications for the treatment of water with UV disinfection:

(1) UV is now perceived as a valuable and viable technology for water disinfection.
(2) Potential costs for meeting upcoming U.S. Federal rules could be substantially lowered by replacing or supplementing ozone with UV disinfection.
(3) The equation for balancing microbial and chemical risks associated with UV treatment must be determined.
(4) Utilities are asking the question: "what else can UV do for water treatment"?

Given the relative understanding of the microbial inactivation capabilities of UV light, the purpose of the proposed research was to further investigate the chemical risks (and potential benefits) associated with implementation of emerging UV technologies.

The overall goal of the research was to examine the effect of the application of UV technologies (with fundamental differences in output spectrum, high and low irradiance, and continuum of energy output) and UV advanced oxidation applications on specific water pollutants such as taste and odor causing chemicals, organic contaminants (e.g., pesticides/herbicides/volatile organic compounds (VOC)) and disinfection byproducts formation (trihalomethones - THMs, haloacetic acids - HAAs, aldehydes, carboxylic acids).

The specific objectives of the research were:

1) Evaluate the efficacy of different UV technologies (LP, MP, P-UV) for treatment of different classes of contaminants in waters of different quality at the bench scale.
2) Evaluate different UV technologies with additions of hydrogen peroxide (H_2O_2) for treatment of different classes of contaminants at the bench scale.
3) Determine the extent of disinfection byproduct formation including aldehydes, carboxylic acids and post-chlorination DBPs, following UV treatment with different UV technologies.
4) Evaluate the relative rates of contaminant destruction using UV and the UV AOP.
5) Pilot test two UV technologies alone and under AOP conditions at two utility field sites for selected contaminant treatment to develop operational experience and preliminary cost and design data.

KEY PROJECT QUESTIONS

This research focused on a number of Key Questions that formed the basis of our experimental plan. These included:

- How do the UV doses necessary for contaminant degradation compare to those needed for UV disinfection?
- How does water quality affect UV contaminant degradation treatment processes?
- How does UV light from different sources affect treatment of different contaminants?
- To what extent does UV-based advanced oxidation improve treatment for various contaminants compared to direct UV photolysis?
- Will UV light form or affect the formation of disinfection byproducts?
- What are the relative efficiencies and costs for treating different contaminants with UV light?

PROJECT TEAM

This project was a joint effort of a number of different research groups under the coordination of Professor Karl G. Linden at Duke University. Duke University was directly responsible for the carrying out the bench scale UV work for all the contaminants and each UV technology. The laboratory work at Duke was led by Dr. Charles Sharpless. The bench work was initially supported by Dr. Jim Bolton and Dr. Mihaela Stefan at Bolton Photosciences, Inc. (BPI) but midway through the project, Bolton Photosciences, Inc. shut down their laboratory and Dr. Stefan moved to Trojan Technologies, Inc. where she continued to assist with the project. Dr. Susan Andrews at the University of Waterloo performed all the DBP testing and analysis for the bench and pilot work in addition to coordinating the pilot work in Kitchener, Ontario Canada. Dr. Mel Suffet at the University of California Los Angeles (UCLA) provided analytical expertise and statistical analysis for the assessment/investigation of the taste and odor contaminants. Dr. Khalil Atasi at Tetra Tech MPS supervised the statistical analyses and Mr. Vinod Korategere coordinated the pilot-scale work in Detroit. Dr. Stefan at Trojan Technologies, Inc. assisted with the data analyses and analytical support for the pilot studies in Detroit, Michigan.

CHAPTER 2
MATERIALS AND METHODS

EXPERIMENTAL SETUP

Low-Pressure Ultraviolet Lamp Units for Collimated Beam Studies

At Duke, the low-pressure mercury (LP) lamp unit was constructed using four 15-W germicidal lamps (ozone-free, General Electric #G15T8) mounted parallel in a reflector housing. The spectrum of the emitted light is given in Figure 3.2. The lamps are seated approximately six inches above an aperture, which is approximately five inches in diameter. A piece of cardboard over the aperture served as a shutter, and users were shielded from the UV emitted by the lamps by another piece of cardboard placed on the side of the reflector housing. For all experiments, the samples were situated so that the solution surface was approximately 30 cm from the aperture. In parallel work at BPI, a 12.4 W LP lamp was placed in the Calgon Carbon unit.

Medium-Pressure Ultraviolet Lamp Units for Collimated Beam Studies

At Duke, the medium-pressure mercury (MP) lamp unit was a Calgon Carbon (Pittsburgh, PA) benchtop instrument. It was used in conjunction with a single 1-kW lamp (ozone-free, Hanovia #6806A441). The spectrum of the emitted light is given in Figure 3.2. The lamp was seated approximately two inches above an aperture approximately three inches in diameter. The instrument housed a pneumatic shutter, and users were shielded from the emitted UV by virtue of the lamp placement in a sealed, ventilated compartment. For all experiments, the samples were situated so that the solution surface was approximately 30 cm from the aperture. At BPI, the MP unit was also a Calgon Carbon instrument, and had a virtually identical design. The light beam was collimated through the 'UV collimator', which has a black-painted roughed-interior surface. The apparatus was provided with both a short and a long collimating tube.

Pulsed-Ultraviolet Lamp Units

Two pulsed-UV (P-UV) lamps were tested, one manufactured by Innovatech (San Diego, CA) and one manufactured by Pulsar (Rosedale, CA). The Innovatech instrument was tested at the Metropolitan Water District of Southern California (MWD). Experiments were conducted at a pulse rate of 10 Hz. The Pulsar unit was tested at Duke. The energy delivered by the Pulsar unit was variable and was operated at 2 Hz at a nominal pulse energy of 352 J, which corresponds to a lamp voltage of 1350 V.

Bench-Scale Reactor

For bench-scale tests of the electrical energy required to degrade contaminants (E_{EO}, see Chapter 6), a closed batch reactor was constructed and used at Duke University (Figure 2.1). This is an annular reactor made from stainless steel and coated with nickel plating. It is capable of housing either one MP or LP lamp in a quartz sleeve (GE type 214) centered in the reactor.

The outer wall of the reactor is hollow to accommodate cooling water from a re-circulating water bath. The temperature was monitored by a thermometer inserted through a vacuum-tight port in the top. The physical light pathlength (distance from the quartz tube to the inner reactor wall) is 10 cm, the reactor height is 17 cm, and the internal volume is very close to 8 L. Stirring was accomplished with an impeller attached to the reactor via a vacuum-tight bearing adapter on top. Contaminants and oxidants were added via syringe through a septum in the top reagent port, which was also used for sampling (via syringe). The lamps used in the reactor differed from those used in the collimated beam work. The MP lamp was 500 W. It used un-doped quartz tubing, and had an emission spectrum closely resembling that of the 1 kW lamp described above. Both lamp and power supply were donated by Hanovia Ltd., U.K. The LP lamp is a 9 W germicidal lamp (GPX 9, Sankyo Denki) designed for use with a ballast that screws into a regular light bulb socket. To seat the lamp in the reactor, wires were soldered from the lamp leads to the ballast, which was screwed into a standard lamp fixture. For all experiments, the lamp intensity was monitored with a radiometer through the quartz window in the side of the reactor. Two different shutters were used depending upon which lamp was installed: for the MP lamp, a stainless steel tube precision mounted on a guide was manually raised or lowered, while for the LP lamp the lamp itself was mounted on a 1" PVC tube that was raised and lowered into the reactor.

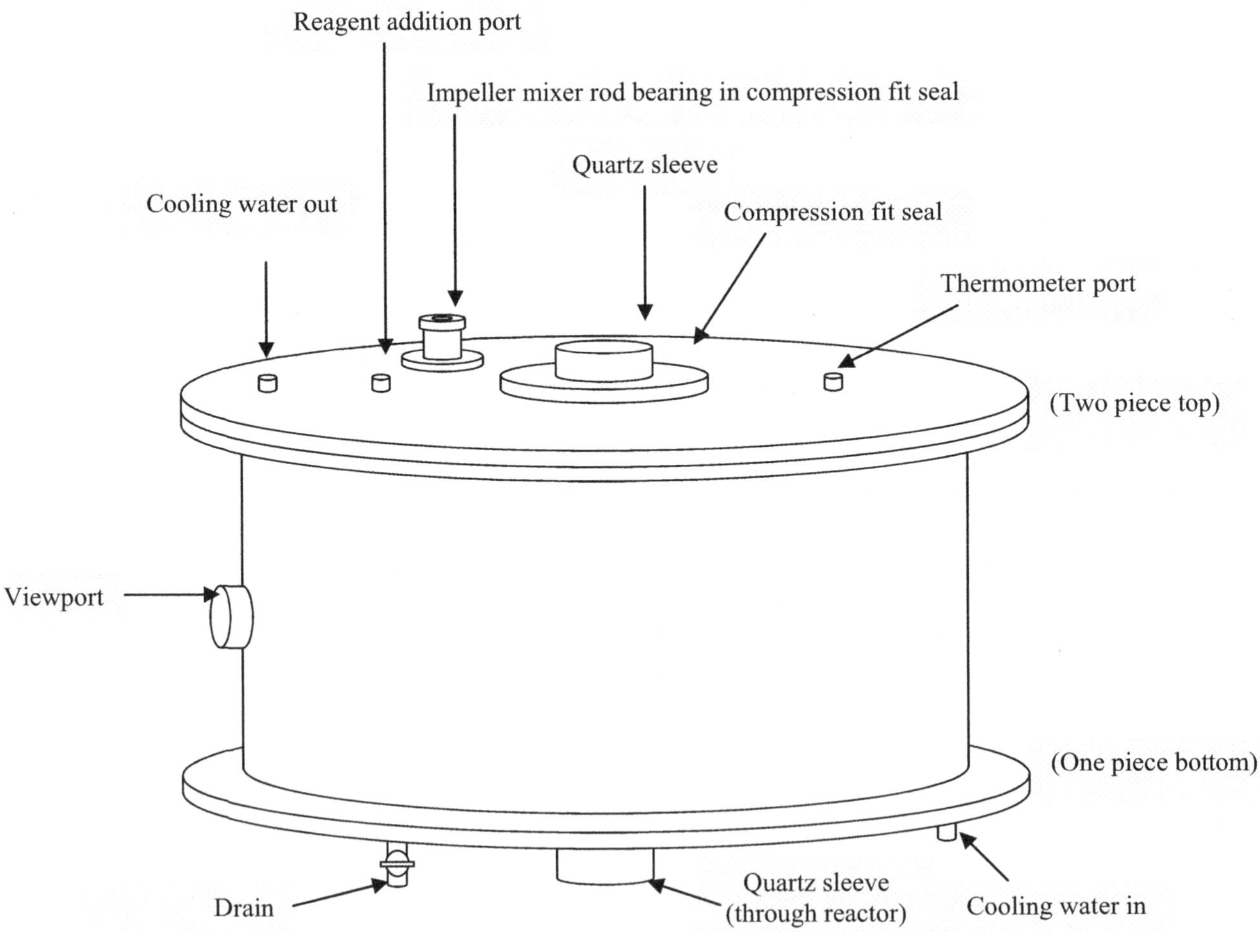

Figure 2.1 Annular reactor design for UV experiments. The reactor has a double wall through which cooling water is circulated

UV Pilot Reactors at Detroit Water and Sewerage Department

Pilot studies were performed at the Southwest Water Treatment Plant, owned and operated by the City of Detroit Water and Sewerage Department (DWSD). The pilot plant consisted of two pilot UV reactor units – a low-pressure lamp UV reactor unit (Trojan 8AL20 UV reactor, manufactured by Trojan Technologies, Inc. London, ON, Canada) and a medium-pressure lamp UV reactor unit (Calgon 4 x 1 kW Sentinel reactor, manufactured by Calgon Carbon Corporation, Pittsburgh, PA, U.S.). A schematic of the UV pilot plant is presented in Figure 7.2, while schematics of the specific reactors are presented in Figures 7.5 and 7.6.

UV Pilot Reactor at Mannheim Water Treatment Plant

The pilot-scale UV reactor located at the Mannheim Water Treatment Plant, Kitchener ON, Canada was manufactured by Calgon Carbon Corporation (Markham, Ontario). It consists of a 111 L (29.4 gallons) UV reactor equipped with six 1-kW Rayox® medium-pressure UV lamps mounted horizontally across a tower-type reactor (Figure 2.2). The UV reactor was installed toward the end of the treatment train, after the biological filters but before the final disinfection, where chlorine or chloramine is applied (Figure 2.3). All six lamps may be set on low or full power levels, and 1 to 5 lamps may be switched off to vary the UV dose from 0 to 6,000 mJ/cm^2. The maximum flow rate through the UV reactor was 200 gpm (gallons per minute), and could be adjusted to as low as 20 gpm at which the maximum UV dose can be achieved according to the manufacturer. The corresponding residence time in the reactor was from 8.3 to 83 s. Reagents (e.g. humic acid, alginic acid, hydrogen peroxide) were injected upstream of a static mixer using chemical pumps.

The Mannheim Water Treatment Plant (MWTP) is a conventional water treatment plant with ozonation and granular activated carbon and dual media filters. The combined conventional and advanced treatment processes are shown in Figure 2.3. Under normal operating conditions, the MWTP treats 27,300 m^3/d (7.2 mgd).

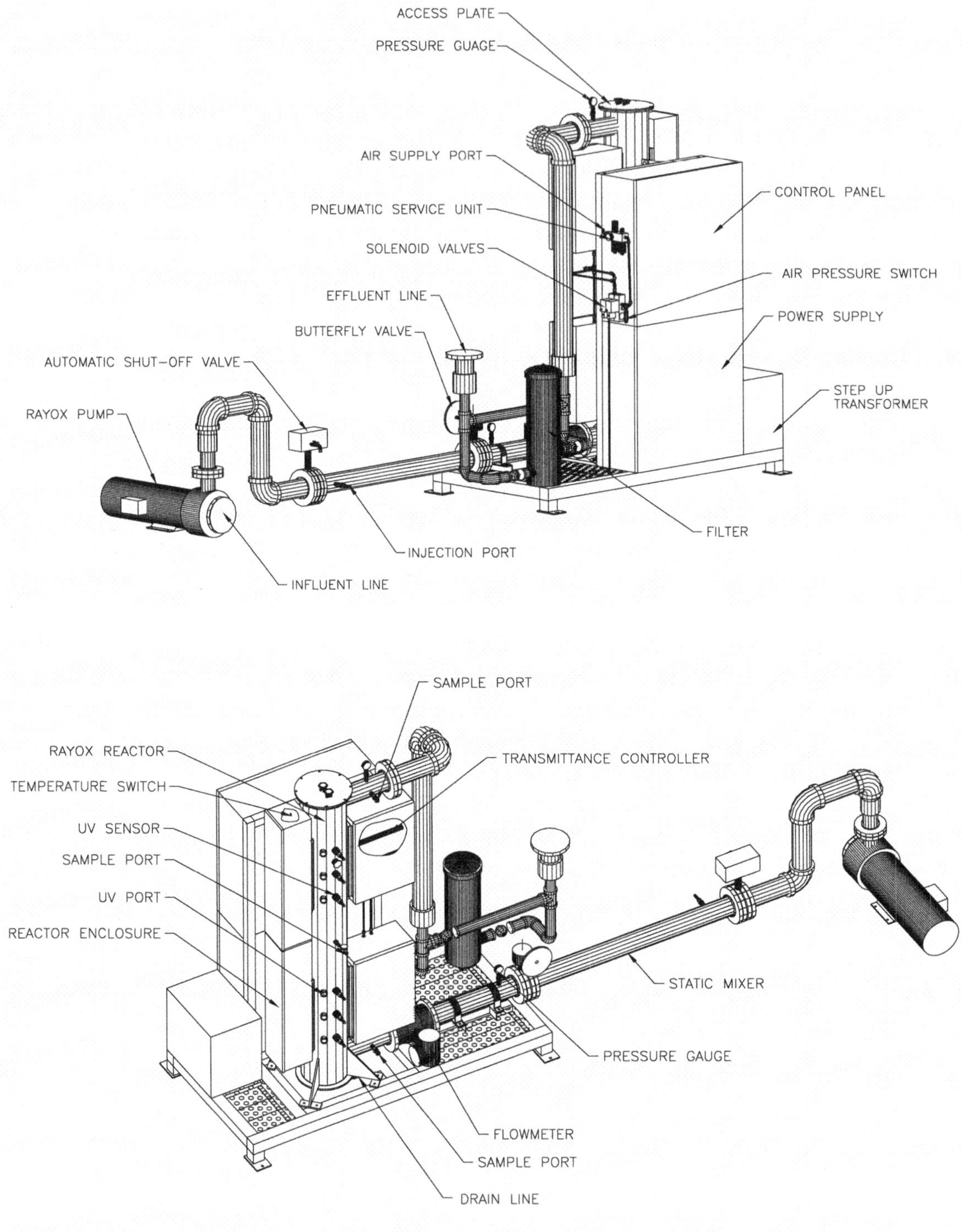

Figure 2.2 Pilot-scale UV reactor at Kitchener ON: (top) front view (bottom) rear view
(Courtesy of Calgon Carbon, Corp.)

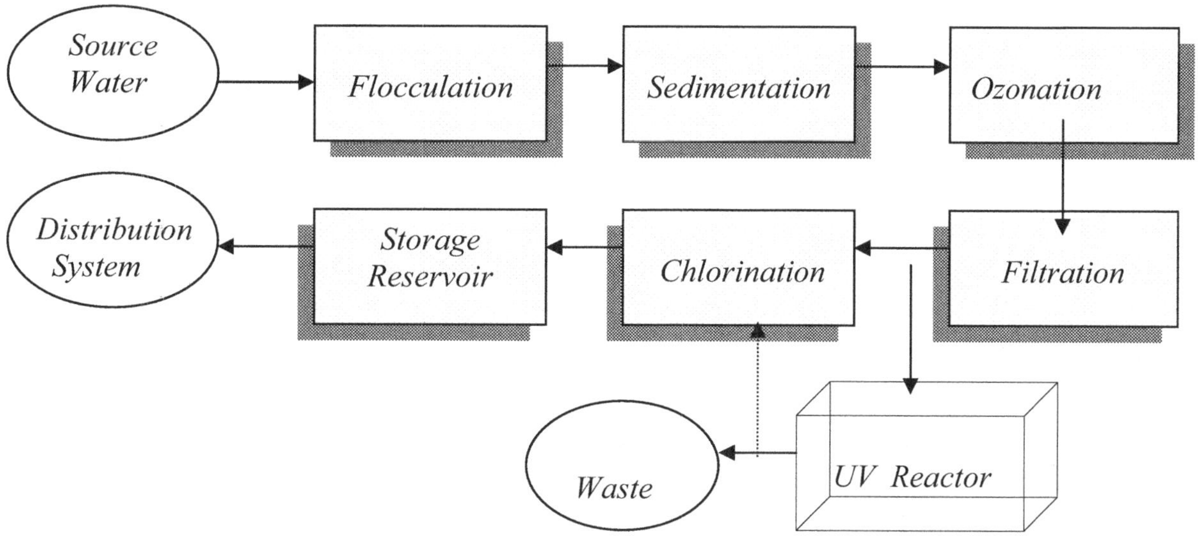

Figure 2.3 Flow diagram of the Mannheim Water Treatment Plant, Kitchener, Ontario

DOSE MEASUREMENT METHODS

Ultraviolet Dose Measurement

The delivered UV dose (mJ/cm^2) over the 200 to 300 nm range was calculated as the average 200 to 300 nm irradiance multiplied by the exposure time. Average irradiance was calculated from the incident irradiance, the normalized lamp emission spectra, and the solution absorbance by standard means (Morowitz 1950). Incident irradiances were determined by chemical actinometry (see below) using identical geometry and solution volumes as in the treatment experiments. Lamp emission spectra for use in irradiance calculations were measured with a diode array spectrophotometer (Ocean Optics, Dunedin FL) whose response was calibrated using a NIST traceable standard D_2 lamp (Thermo Oriel, Stratford CT); the spectra were then normalized from 200 to 300 nm. In some cases, the solution absorbance decreased during the course of the experiments; this was continuously monitored and the data were used to determine the average UV dose delivered to the solution at sampling times by assuming the average irradiance between two samples to be equal to the mean of the irradiances at those points.

Chemical Actinometric Methods

Incident irradiance values (mW/cm^2) over the 200 to 300 nm range were measured in separate experiments by chemical actinometry in conjunction with spectral measurements. Chemical actinometry was used to quantify the incident photon irradiance (einstein cm^{-2} s^{-1}). Two actinometers were used: iodide/iodate actinometry (Rahn 1997) for the LP lamp, and ferrioxalate actinometry (0.006 M $K_3Fe(Ox)_3$) (Murov et al. 1993) for the MP and pulsed lamps. Actinometric photon irradiances were correlated to radiometer readings (Model IL1700, International Light, Newburyport, MA) and the radiometer was used to adjust the actinometry results for daily fluctuations in the lamp intensity. With LP, the irradiance (mW cm^{-2}) was simply calculated from the photon irradiance (determined from the rate of I_3^- production, (Rahn 1997)) and the photon energy at 254 nm. For MP and pulsed lamps, the irradiance was defined as the energy delivered below 300 nm. The photon irradiance in this range was calculated from the difference between the Fe^{2+} production rates with and without a 3 mm thick 295 nm long-pass filter (Schott, WG295) placed in the light path. A detailed account of the procedures and calculations involved in this method can be found elsewhere (Sharpless and Linden 2003). For the pulsed lamps, actinometry was used to determine the incident energy flux per pulse (averaged over 20-30 pulses), which allowed calculation of the number of pulses necessary to achieve the desired UV doses.

Bench-Scale Reactor Dose Measurement

For experiments in the bench-scale reactor, UV dose rates were estimated from the decay rate of NDMA in the test waters (synthetic water and Durham filtered water). Because the goal of these experiments was to determine relative E_{EO}'s of the target compounds, it was not necessary to accurately quantify the delivered UV dose, only the reaction times and power consumption, whose measurement is described in Chapter 6. The estimated dose rates based on NDMA degradation were used to determine appropriate experimental exposure times for the other contaminants.

SOURCE WATER QUALITY

Laboratory Waters

In several experiments, water was made in the laboratory to simulate natural surface water. This water is referred to as "synthetic water 1" and was prepared using $CaCl_2*2H_2O$, $MgCl_2*6H_2O$, $NaNO_3$, $CaSO_4*2H_2O$, $NaHCO_3$, and two types of natural organic matter (NOM), Suwannee River NOM (International Humic Substances Society, www.ihss.gatech.edu) and Alginic Acid (Sigma-Aldrich). The water, which had a pH of approximately 8.3 and total organic carbon concentration (TOC) of 3 mg/L, was prepared from various stock solutions within one week of performing the experiments. Its ionic composition is given in Table 2.1. This water had a hardness of approximately 110 mg/L as $CaCO_3$ and an alkalinity of approximately 75 mg/L as $CaCO_3$. Its absorbance spectrum is shown in Figure 2.4.

Table 2.1
Composition of synthetic water 1

	Ca^{2+}	Mg^{2+}	Na^+	$(CO_3)_{TOT}$	NO_3^-	Cl^-	SO_4^{2-}	NOM	Alginic Acid
mg/L	29.1	9.99	36.2	90	3.00 (0.679 as N)	40.0	55.0	2.56	5.32

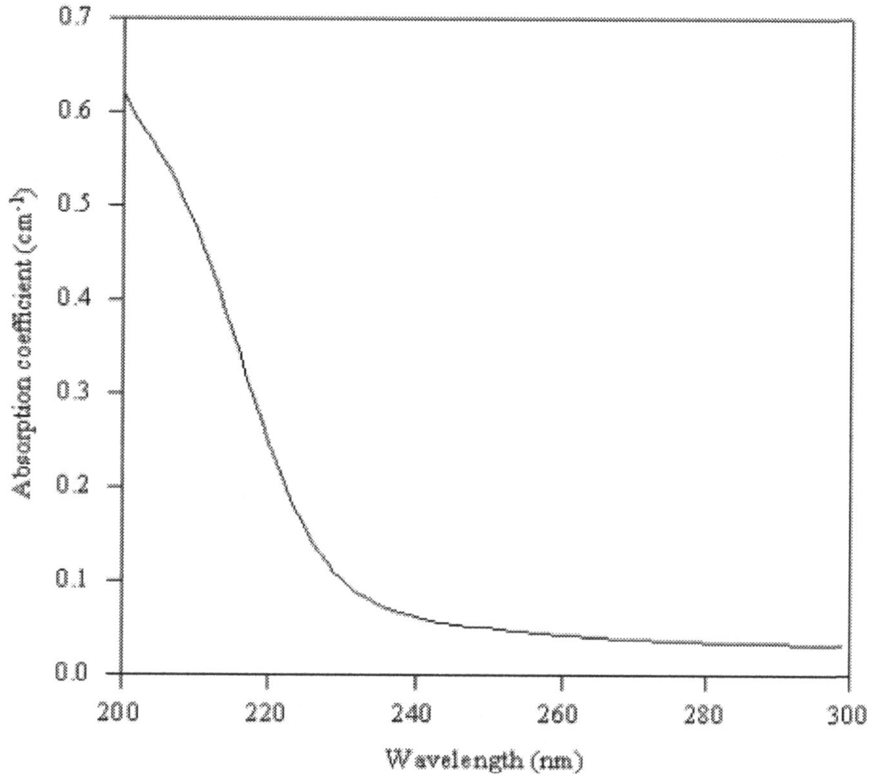

Figure 2.4 Absorbance spectrum of synthetic water

For bench-scale experiments on water quality effects it was necessary to vary the pH, alkalinity, and TOC simultaneously. This was accomplished by eliminating the Ca and Mg salts from synthetic water 1, incorporating phosphate into the recipe as a buffer, using NaCl to match the ionic strength of synthetic water 1, and varying the concentration of NOM. The concentration of $NaNO_3$ remained unchanged. The UV absorbance spectra of these waters were quite similar to that of synthetic water 1 except that the absorbance above 230 nm increased as TOC levels increased. The waters prepared in this manner used varying amounts of phosphate to achieve a balance between buffering and maintaining constant ionic strength and had a significantly lower hardness than synthetic water 1. As a check that this change would not substantially affect the experimental results, control experiments were performed on atrazine (ATZ) at BPI. Similar results were obtained at pH 8.3 using both synthetic water recipes.

Natural Waters

Durham Filtered Water

For E_{EO} measurements using the bench-scale reactor at Duke, filtered water was obtained from the Williams Plant on Hillandale Road in Durham, NC. This is a 22 mgd plant, which takes its water from two lakes, Lake Michie and Little River Lake. The water is chlorinated upstream of filtration to prevent bio-fouling. The experimental water was taken from filter number 4, and the pertinent water quality parameters are given in Table 2.2.

Table 2.2
Water analysis for Durham Filtered Water

pH	Alkalinity (mg/L as $CaCO_3$)	Hardness (mg/L as $CaCO_3$)	Nitrate (mg/L as N)
6.8	8.0	32	0.4

Southwest Water Treatment Plant

Pilot studies were performed at the Southwest WTP, one of the five water treatment plants owned and operated by the City of Detroit Water and Sewerage Department (DWSD). The plant has a rated capacity of 180 mgd, and gets its raw water from the Detroit River through an intake off of Fighting Island. The treatment process consists of pre-chlorination and hydrofluosilicic acid addition (for fluoride) at the low lift caisson, liquid alum addition (primary coagulant) and powdered activated carbon addition (used seasonally only for taste-and-odor) during rapid mix, flocculation, sedimentation, dual-media filtration using anthracite/sand, followed by post-chlorination and phosphoric acid addition (for corrosion control).

The Southwest WTP finished water was used as feed water for the UV pilot plant. A 3-inch line that provides flushing water to the raw water screens (at approximately 70 psi pressure) was used as the feed water source. A summary of the Southwest WTP finished water quality during this study is presented in Table 2.3. The finished water has very low turbidity, low TOC, neutral to slightly basic pH, moderate alkalinity and hardness, and very low iron and manganese.

Table 2.3
Southwest WTP Finished Water Quality

Water Quality Parameter	Average	Range
Turbidity (ntu)	0.07	0.05 – 0.10
pH	7.35	7.3 – 7.5
Alkalinity (mg/L as $CaCO_3$)	72	70-76
Total Hardness (mg/L as $CaCO_3$)	106	95-110
Iron (mg/L)	< 0.05	< 0.1
Manganese (mg/L)	< 0.01	< 0.01
TOC (mg/L)	1.5	1.4 – 1.7

Mannheim Water Treatment Plant

The raw water source for the pilot-scale UV reactor located at the Mannheim Water Treatment Plant, Kitchener ON, is the Grand River, a water with high alkalinity and moderate organic level. The typical water quality data for this source water and finished water are presented in Table 2.4.

Table 2.4
Typical water quality data for the Grand River and Mannheim WTP finished water

Parameter	Grand River *	Finished Water **
Temperature, °C	0.1 - 25.8	13.0
pH	6.7 - 7.7	7.4
Alkalinity, mg/L as $CaCO_3$	150 – 200	224
Magnesium, mg/L	18 – 30	< 0.01
Calcium, mg/L	40 – 100	–
Hardness, mg/L as $CaCO_3$	210 – 300	290
Color, true, TCU	1 – 2	1
DOC, mg/L	2.0 - 3.6	1.8
Turbidity, NTU	< 1	< 0.1
Total Ammonium, mg/L	0.00 - 0.09	–
Nitrate, mg/L	0.9 – 2.9	3
Sulfate, mg/L	48 – 60	52
Fecal coliforms MF, /100mL	10 – 550	0
TTHMs, µg/L ***	_	17.6
HAA_9 , µg/L ***	_	23.6

* Adapted from Anderson et al. 1994
** Adapted from Water Quality Data Report for the Water Supply System of the Regional Municipality of Waterloo, 1997
*** After chlorination and subsequent ammoniation to form chloramines

ANALYTICAL METHODS

Standard Methods

Total organic carbon (TOC) measurements were made on a Tekmar-Dohrmann (Cincinnati, OH) Apollo 9000 carbon analyzer. Hydrogen peroxide was determined by the tri-iodide method (Klassen et al. 1994). A glass electrode that is calibrated daily was used to measure pH. Alkalinity was determined by titration (APHA 2000, Standard Methods, 2320). Nitrate was determined by colorimetry (APHA 2000, Standard Methods, 4500). Ultraviolet light absorbance was measured on a Cary Bio 100 spectrophotometer (Varian, Houston, Texas) in a 1-cm pathlength cuvette.

Methods for Contaminants

At Duke, all contaminants were analyzed by reversed-phase HPLC using a Varian ProStar system (Varian Scientific Instruments, Inc.), comprising two pumps, an autosampler, a photodiode-array UV absorbance detector, and a 150 x 4.6 mm C-18 cartridge column (Chrompack, Varian Inc.). For NDMA, the mobile phase was 15% MeOH : 85% 25 mM phosphate buffer at pH 6.5 (1 mL/min), detection was at 230 nm, and with a 50 µL injection the detection limit was approximately 10 nM (0.7 µg/L). For ATZ, the mobile phase was 1:1 acetonitrile:water (1.0 mL/min) detection was at 220 nm, and with a 50 µL injection the detection limit was approximately 5 nM (1 µg/L). In bench-scale reactor tests, a mixture of ATZ, carbaryl, toluene, metolachlor, and diazinon was analyzed with a mobile phase of 65%

aqueous acetonitrile (1 mL/min) and detection at 230 nm (for ATZ and carbaryl), and 210 nm for the other contaminants. With a 100 μL injection the detection limit ranged from 0.4 nM (for carbaryl) to 8 nM (for toluene).

At BPI, atrazine concentration was measured by a HPLC method. A Waters HPLC system equipped with a Waters 600 pump, a Waters 717 autosampler and a Waters 996 photodiode array detector was used with a Supelco Suprasil C-18 reversed phase analytical column (250 mm × 4.6 mm, 5 μm particle size), and a mobile phase [acetonitrile: 5 mM phosphate buffer (pH 6.5), 1:1 ratio (vol/vol)] at a flow rate of 1.2 mL/min. The detection wavelength was set at 228 nm. Under such conditions, atrazine concentrations as low as 1 ppb (μg/L) can be accurately measured.

MIB, IPMP and geosmin (GSM) were analyzed by using solid-phase microextraction (SPME) and gas chromatography-mass spectrometry according to *Standard Methods* (APHA et al., 2000).

The taste and odor (MIB and geosmin) compounds from the pilot studies were analyzed at Trojan Technologies Inc. Analyses were performed with a Varian 3800 Gas Chromatograph with a Saturn 2000R Mass Spectrometer (GC/MS) equipped with a CombiPal (CTC Analytics Inc.) autosampler to analyze MIB and GSM. The analytical procedure followed a modified Supelco Application Note 147, based on the headspace solid-phase microextraction (SPME) technique. MIB and GSM were extracted onto the SPME fiber (2 cm StableFlex coated with 50/30 μm DVB/Carboxen/PDMS, Supelco) exposed for 20 min to the headspace of the 20 mL autosampler capped-vial. After extraction, the fiber was immersed into the GC injection port (250°C) where the analytes were desorbed for 3 min. A temperature program was run through a 30 m x 0.53 mm, 1 μm coating thickness GC column: 40 °C (3 min hold) to 250°C at 40°C /min (5 min hold), using helium as a carrier gas. The quantitation ions were 95 and 112, for MIB and GSM, respectively. Under these conditions, MIB and GSM concentrations as low as 5 ppt (ng/L) can be accurately measured.

DBP Methods

DBP analyses and Uniform Formation Condition (UFC) tests were initiated upon receipt of the samples at the UW laboratory. Aldehyde concentrations were determined by GC-ECD analysis of extracted PFBHA derivatives as per *Standard Methods* (method 6252, APHA, 2000). Carboxylic acid concentrations were determined by ion chromatographic analysis of chloroform-preserved samples (Peldszus et al., 1996). UFC tests were performed according to the procedure reported by Summers and co-workers (1996). Residual chlorine was quenched with sodium thiosulfate immediately at the conclusion of the 24-hour reaction period, following which THM and HAA analyses were performed according to *Standard Methods* (method 6232 and 6251for THMs and HAAs, respectively, APHA 1998) with minor modifications. When H_2O_2 was present in the samples, 0.2 mg/L catalase was added immediately after irradiation to quench the residual H_2O_2. Experiments performed at the University of Waterloo (Appendix A) showed that this level of catalase does not contribute significantly to the level of DBPs, nor does it noticeably increase the TOC concentration. Travel standards and travel blanks for each class of DBPs were included with each batch of samples. Laboratory standards and blanks were also prepared at the time of analysis of each set of samples.

CHAPTER 3
COMPARISON OF UV TECHNOLOGIES FOR DEGRADATION OF NDMA AND ATRAZINE

INTRODUCTION

While UV is fully established as an effective method for inactivating water-borne microbial pathogens, its potential for removing chemical contaminants is just beginning to be realized. This potential has been known for many years, but application of the technology for general water treatment purposes had to wait until it was widely available and could be easily integrated into existing treatment plants. Over the past few years, this scenario has come to pass, due principally to two factors. First, the U.S. EPA's recent emphasis on tighter regulation of chlorinated disinfection by-products has led utilities to search for alternative disinfectants. Second, the new-found awareness of how susceptible certain microorganisms (e.g., *C. parvum*) are to UV disinfection has led to rapid development of a wide array of UV systems for drinking water treatment.

One of the principle challenges for utilities in the coming years will be to figure out how to achieve multiple water quality goals with a minimum of capital investment. Thus, it is important to understand the scope and limitations of UV as a treatment method for chemical contaminants. At present, there are certain chemicals that are clearly a concern for utilities; many of these chemicals are on the Candidate Contaminant List assembled by the U.S. EPA. Others, such as N-Nitrosodimethylamine (NDMA), have been identified as treatment priorities in Canada and several U.S. states. Of the many potential target compounds, two stand out due to their known susceptibility to UV and UV/H_2O_2 processes. These are NDMA and ATZ.

In this work, bench-scale studies were conducted in simulated natural water to assess the effectiveness of different UV lamps, and those lamps combined with H_2O_2 addition, for treating NDMA and ATZ. One of the more important results to come out of this study is a better understanding of how to compare bench-scale results using different UV lamps and experimental set-ups, and the necessary considerations are described below. In many cases, the applied UV dose can be a meaningful descriptor of process efficiency, but it is critical to understand how the UV dose is defined and measured and how it relates to the particular lamp emission spectrum.

BASIC CONSIDERATIONS FOR COMPARING UV TECHNOLOGIES

It is impossible to compare UV technologies without reference to a specific application. In the case of chemical oxidation or contaminant photolysis, the application requires overlap of the UV lamp emission spectrum with either or both the absorption spectrum of the contaminant or H_2O_2 (for the UV/H_2O_2 process). The so-called First Law of Photochemistry dictates this requirement: only light that is absorbed can cause a chemical change. If the light is not absorbed, the energy is simply wasted. Competition between the contaminant and natural water constituents can also be an important consideration. In the UV/H_2O_2 process, the absorption spectrum of the contaminant sometimes overlaps that of H_2O_2; in this case the competition for photons is between the contaminant and the oxidant and it will need to be considered when trying to optimize the process efficiency. The following sections describe these considerations with respect to NDMA and ATZ with the lamps used in this study.

Contaminant Absorption Spectra and Lamp Emission Spectra

Contaminant absorption spectra

Both NDMA and ATZ strongly absorb UV radiation below 300 nm, and the absorption spectra of these compounds in water are shown in Figure 3.1. For NDMA, the critical wavelengths are in the 200 to 275 nm wavelength range, while for ATZ they cover the whole range below 300 nm. Clearly, any UV source with good output below approximately 270 nm should be useful for photolyzing both contaminants.

Also shown in Figure 3.1 is the absorption spectrum of H_2O_2 (magnified 100x). Its absorbance overlaps that of both NDMA and ATZ, indicating that besides its role as an ·OH source it can also screen light from the contaminants and may either slow down or accelerate their photolysis depending upon the amount of H_2O_2 applied and the rate at which the contaminant reacts with ·OH.

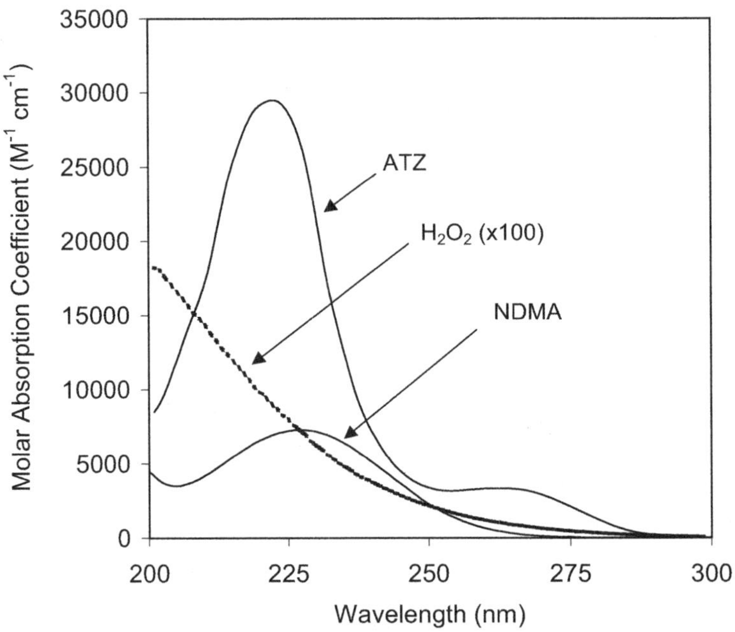

Figure 3.1 Molar absorption spectra of ATZ, NDMA, and H_2O_2 (x 100)

Lamp Emission Spectra

LP UV lamps emit essentially monochromatic light at 254 nm, while MP UV lamps have various outputs ranging from about 205 nm to above 500 nm. The physical characteristics of these lamps were described in Chapter 1, and their emission spectra are shown in Figure 3.2. With regard to process efficiency, besides the range of wavelengths emitted, the critical differences between the two lamps are the emitted UV intensity, or irradiance (higher for MP), power consumption (higher for MP, energy losses in heat), and electrical efficiency (higher for LP).

Pulsed UV lamps also appear to be quite effective for treating NDMA (Sharpless et al., 2001). These light sources are distinctly different from the mercury vapor-based lamps;

typically, a noble gas or gas mixture is used, and an electrical discharge produces a brief (μs duration) but intense pulse of light. The spectral distribution is approximately that of a blackbody (Figure 3.3), where the peak wavelength depends on the temperature, and it can be adjusted by varying the gas mixture, pressure, and applied voltage.

All three types of lamps emit UV in the critical range of 200 to 300 nm where NDMA and ATZ absorb light. Thus, they will be effective at degrading these contaminants. For utilities, the question is how efficient each lamp type will be with regard to power consumption per amount of contaminant removed. How this can be assessed on the bench-scale in collimated beam experiments is not immediately obvious, but as discussed below it is possible to estimate their relative efficiencies based on the UV dose required for degrading the contaminants.

Comparisons Based on Time Versus those Based on UV Dose

When trying to compare different UV lamps, it might seem ideal if all the lamps in question could simply be swapped in and out of a given reactor. However, this approach neglects the fact that different reactor designs are optimal for different lamp types. As a simple example, consider LP and MP lamps; the lower irradiance provided by LP lamps requires them to be more closely spaced than MP lamps in a reactor to achieve a desired irradiance in the reactor. The collimated beam approach used here also suffers from the disadvantage that several substantial differences exist between the experimental configurations for the different lamps. The LP system houses 4 lamps while the MP and pulsed systems only house one lamp, and there are differences between the sizes of the apertures on the housings for the different lamps. Also, the solutions were placed at slightly different distances from the different lamps. These physical parameters exert great control over the reaction rates making it impossible to use time-based rate constants to compare the lamps.

These considerations illustrate a general problem with reporting time-based rate constants for photochemical reactions: unless a great many variables are specified, such "constants" are not a useful way to quantify reaction rates. It is more appropriate to report volume or area normalized photon- (einstein) or energy- (mW) based rate constants or to report the apparent quantum yield. Another approach is to consider the average UV dose-based rate constant for comparing the rates. The units on these rate constants are inverse UV dose, or $cm^2\ mJ^{-1}$, and the parameter itself quantifies the surface area of a solution over which a given energy may be spread to achieve a certain level of contaminant removal. Thus, higher dose-based rate constants indicate a greater extent of reaction for a given UV dose. As a means of comparing UV lamps, the dose-based rate constant, or k_d', has the advantage that it depends only weakly on the physical geometry of the experiment for polychromatic UV lamps and not at all for monochromatic sources. The biggest factor affecting k_d' for polychromatic lamps is the solution depth because as the depth increases, the apparent lamp spectrum changes due to greater attenuation of the shorter wavelengths. This leads to different wavelengths contributing to the average irradiance at different depths and can change the measure k_d' slightly depending upon the experimental depth. Otherwise, k_d' values are virtually ideal for comparing UV lamps because they are based on the average UV dose, which is corrected for solution transmittance, and because once the UV dose for 1-log removal is specified, that information can be used to determine electricity requirements for a given reactor/lamp combination whose irradiance is known.

Time-based rate constants are, however, useful for determining whether process changes enhance or diminish the electrical efficiency for a given lamp type. For example, if adding H_2O_2 increases the time-based rate constant, or k_t', it clearly enhances the electrical efficiency. The same cannot automatically be said if k_d' increases without specifying how the average irradiance is altered by the addition of the H_2O_2. If increasing k_d' by 10% required adding enough H_2O_2 to diminish the average irradiance by 15%, the exposure time would need to be greater to achieve the required UV dose, and more electricity would be used. Thus, both or either k_d' and k_t' can be meaningful descriptors of the system performance depending upon the desired information: k_d' is more useful for comparing lamp types while k_t' is more useful for examining the effect of process changes.

RESULTS

Spectra of MP, LP, and Pulsed UV lamps

Figures 3.2 and 3.3 show the actual spectra of the lamps used in these experiments. To avoid raising issues related to commercialization potential of the two pulsed lamp units that were studied, they are referred to here as P1 and P2. The main difference between the two lamps is that the peak wavelength appears to be shifted to shorter wavelengths for P2. However, for the interpretations presented here, the crucial difference is that for a given unit of energy delivered (mJ cm^{-2} $pulse^{-1}$) over the 200 to 300 nm wavelength range, the spectral distribution of that energy is more uniform for P1 whereas for P2 it is predominantly from wavelengths longer than 250 nm.

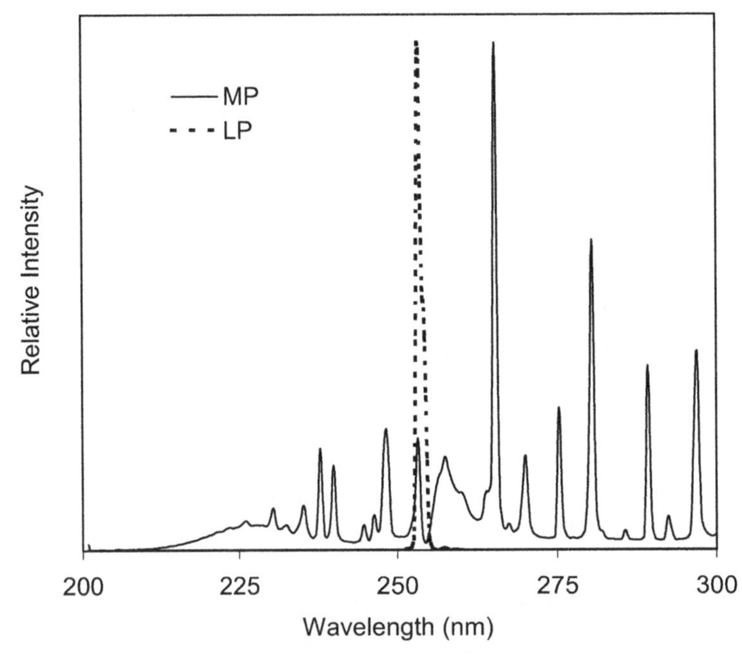

Figure 3.2 200 to 300 nm emission spectra of MP and LP lamps used in this work

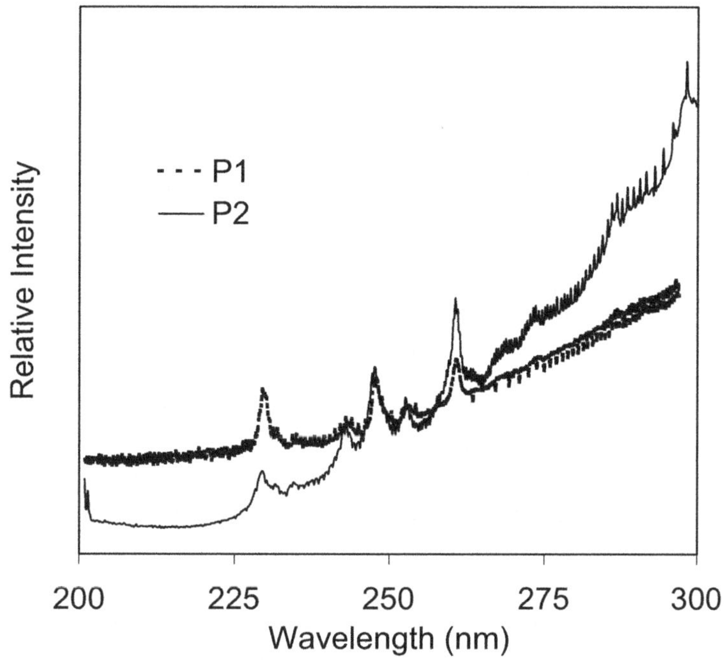

Figure 3.3 200 to 300 nm emission spectra of pulsed lamps used in this work

Kinetic data

All experiments except those at Bolton Photosciences, Inc. (BPI) were conducted with an initial contaminant concentration of 1 μM; at BPI, the initial concentration was 0.5 μM. All experiments were conducted in synthetic water 1, which has a pH of 8.3. Figures 3.4 and 3.5 present pseudo-first order kinetic plots for ATZ and NDMA degradation for all the UV lamps tested in Phase I at Duke, BPI, and the Metropolitan Water District of Southern California (MWD). The rate constants derived from these data are given in Table 3.1. Under all conditions tested, pseudo-first order kinetics were observed, which indicates that these rate constants are applicable to lower concentrations of contaminants such as are found in natural waters. Note that these rate constants are in terms of the natural log of the concentrations according to Equation 3.1:

$$\ln\left(\frac{[\mathrm{C}]}{[\mathrm{C}]_0}\right) = \mathrm{k}_\mathrm{d}^{'} H_{200\text{-}300} \qquad 3.1$$

Here, k_d' is the UV dose-based rate constant ($cm^2\ mJ^{-1}$) and $H_{200\text{-}300}$ is the UV dose from 200 to 300 nm ($mJ\ cm^{-2}$). The value of k_d' thus represents the inverse of the UV dose required for the concentration to drop to 1/e of its initial value. To get the UV dose for 1-log removal, k_d' can be inverted and multiplied by 2.3.

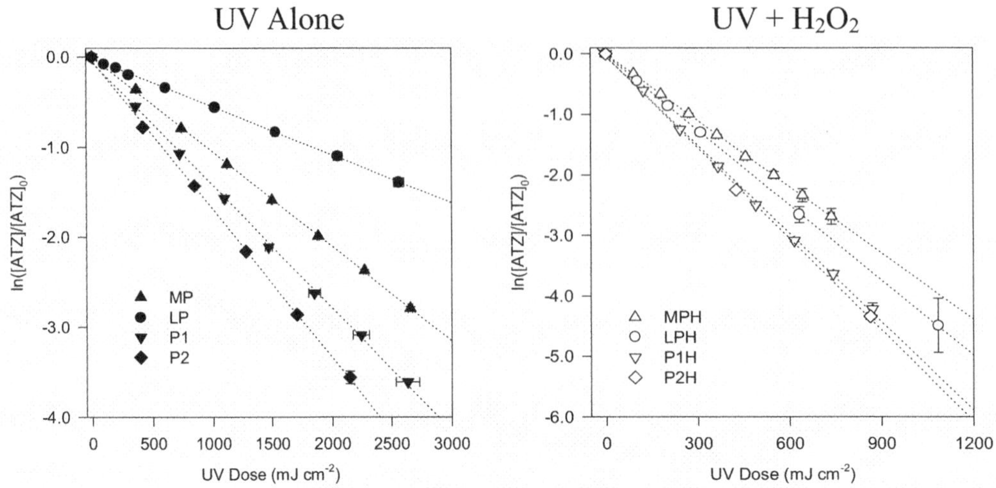

Figure 3.4 Kinetic plots for ATZ photolysis with all lamps. Data for LP and MP from Duke

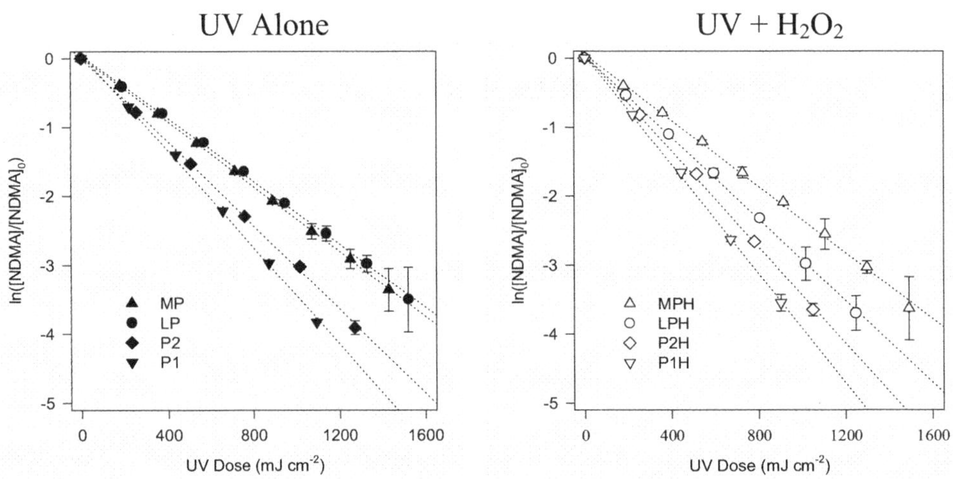

Figure 3.5 Kinetic plots for NDMA photolysis with all lamps

Table 3.1
Rate constants for contaminant removal in dose-based units ($cm^2/mJ \times 10^3$) using the 200 to 300 nm UV dose

	LP	LP/ H_2O_2	MP	MP/ H_2O_2	P1	P1/ H_2O_2	P2	P2/ H_2O_2
ATZ (D)	0.54	4.1	1.1	3.7	1.4	4.9	1.6	5.0
ATZ (BPI)	0.52	3.8	1.5	5.4	-	-	-	-
NDMA (D&MWD)	2.3	2.9	2.4	2.4	3.5	3.9	3.0	3.5

Participating Institution Abbreviations:
D = Duke BPI = Bolton Photosciences MWD = Metropolitan Water District

In general, NDMA is more rapidly destroyed by direct photolysis while ATZ is more rapidly destroyed by photolysis driven advanced oxidation with the addition of 100 mg/L H_2O_2. In fact, only small or no relative increases are observed in the NDMA removal rates with added

H_2O_2, but for ATZ the enhancements are approximately 3- to 8-fold depending on the UV lamp used. This difference is discussed in more detail below.

Comparison of the dose-based rate constants in Table 3.1 shows that more NDMA removal can be achieved per-dose for the pulsed systems than for the LP or MP lamps and that MP and LP appear to be equally effective. For ATZ, the difference between the pulsed and Hg-based lamps is also there but is much smaller than with NDMA. Also, for ATZ, the MP lamp appears to be significantly more effective than the LP lamp in the absence of H_2O_2; the two lamps appear more comparable in the case of the UV/H_2O_2 process. Another thing to note in Table 3.1 is that dose-based rate constants measured at Duke and BPI for ATZ are very similar for the LP lamp, but for the MP lamp there is a significant difference (the value measured at BPI is approximately 36% higher). The source of this discrepancy was traced to differences in the lamps at the two locations, and this is discussed in more detail below.

Cross-Laboratory Study of Atrazine Degradation by MPUV

As noted above, the rate constants for ATZ were measured both at Duke and at BPI. The values obtained with the LP lamps are in excellent agreement, but there is a large discrepancy between the results with the MP lamps. Additional experiments confirmed that this difference was not due to the differences in the initial ATZ concentrations or the UV dose calculations but to differences in the emission spectra of the lamps used at the two locations.

Normalized emission spectra of the two lamps are shown in Figure 3.6. When comparing the spectra in Figure 3.6 with the ATZ spectrum in Figure 3.1, one can see that there is a significantly stronger output of the BPI MP lamp in the 220-265 nm range compared with the output of the Duke MP lamp. This range of wavelengths overlaps a large part of the absorption spectrum of ATZ, and this means that for a given UV dose, more of the energy is absorbed by ATZ for the BPI lamp than for the Duke lamp.

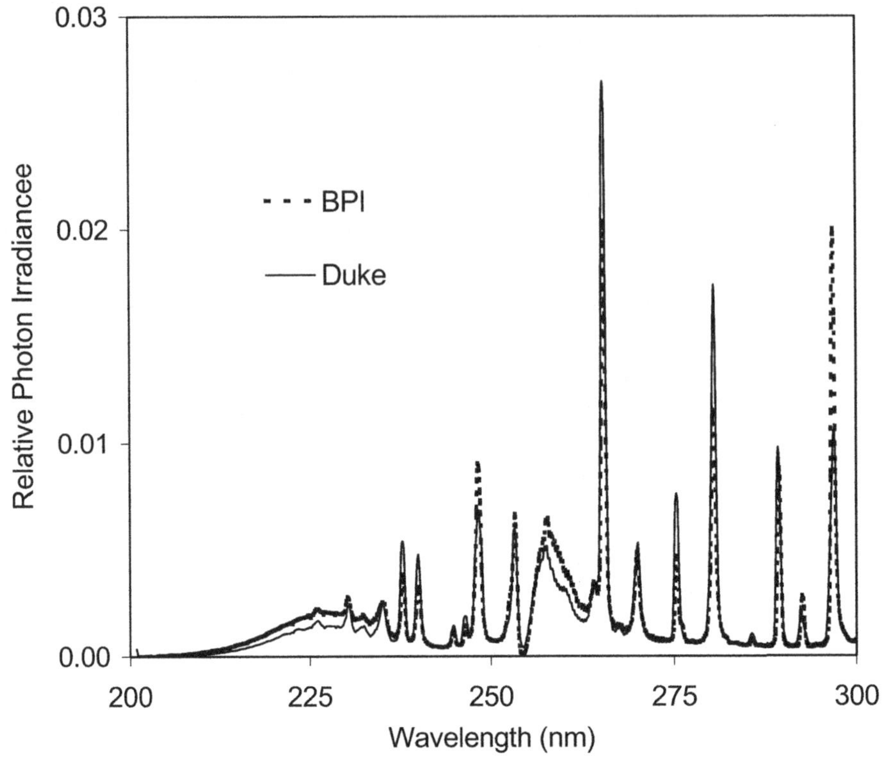

Figure 3.6 Emission spectra of the MP lamps used at Duke and BPI

The spectra were used in combination with experimental data on atrazine photolysis and its absorption spectrum to calculate the rate constants that should be expected from both lamps as described in detail elsewhere (Stefan et al. 2001). The pertinent results that came out of this analysis are that the quantum yield for ATZ direct photolysis is higher in the shorter wavelength absorption band, approximately 0.05, than in the longer wavelength absorption band, approximately 0.03. These results have been double checked with further measurements and modeling (Sharpless et al. 2003), and more accurate values of the quantum yields are believed to be 0.055 and 0.038 for the shorter and longer wavelength absorption bands, respectively. Using these data and standard photochemical theory (Schwarzenbach et al. 1993; Stefan et al. 2001), the theoretical rate constants are calculated to be 1.08 and 1.50 x 10^{-3} cm^2/mJ for the Duke and BPI lamps, respectively. These agree exactly with the measured values within experimental uncertainty. This exercise illustrates how and why different k_d' values can be expected with different polychromatic lamps for certain contaminants, in particular those with multiple absorption bands at wavelengths used to define the UV dose.

DISCUSSION

UV versus UV/H_2O_2

Both NDMA and ATZ are susceptible to direct photolysis by UV alone, with NDMA being the more susceptible. This is borne out by calculation of the quantum yields for photolysis of the two compounds: ATZ has a quantum yield of no more than 0.055 at 220 nm while for NDMA it is approximately 0.30 at 230 nm and pH 8 (Sharpless and Linden 2003; Stefan and

Bolton 2002; Stefan et al. 2002). Although the molar absorption by ATZ is almost three times greater than NDMA in its major absorption band, this difference does not make up for the five- to six-fold difference in the quantum yields. Thus, to achieve 1-log removal of the contaminants with the LP lamp requires $H_{200\text{-}300}$ to be in the range of 4,500 mJ/cm^2 for ATZ but only 1,000 mJ/cm^2 for NDMA; for the polychromatic lamps, these values range from 1,500 to 2,100 mJ/cm^2 for ATZ and 650 to 1,000 mJ/cm^2 for NDMA.

The addition of 100 mg/L H_2O_2 affects the removal of NDMA and ATZ in very different ways, and the magnitude of the effect depends upon the lamp type. As can be seen in Table 3.1, the enhancement of ATZ degradation with added H_2O_2 is substantially larger for the LP than the MP lamp. At both Duke and BPI, the relative enhancement in the rate constant is almost exactly twice as large with LP than MP lamps. In contrast, only small if any enhancements were observed for NDMA by adding H_2O_2. These effects can be understood by considering that NDMA reacts with ·OH almost 10 times slower than does ATZ: the second order rate constants for reaction with ·OH are 3.3×10^8 and 2.6×10^9 M^{-1} s^{-1} for NDMA and ATZ, respectively (NDRL 2003). Add to this the large amount of light screening by H_2O_2 at 100 mg/L, and what occurs is that the relatively efficient direct photolysis of NDMA is slowed by light screening, which is just barely made up for by reaction with ·OH. For ATZ, light screening slows photolysis in the short wavelength absorption band but only slightly in the long wavelength band, and the fast reaction with ·OH leads to a large increase in the observed rate of photolysis.

To assess the effect of H_2O_2, k_d' values may be considered, but this appears to be of limited value for practical purposes. Because the UV dose accounts for light screening by H_2O_2 (i.e, the average irradiance is quite different between runs with and without H_2O_2), it does not relate in an immediately obvious way to the time required for contaminant removal. Thus, it does not indicate how H_2O_2 affects the electrical efficiency of the process. In contrast, comparison of how the time-based rate constant, k_t', changes with H_2O_2 illustrates how the H_2O_2 changes the electricity requirements for the process. Values for k_t' (min^{-1}) are given in Table 3.2. For ATZ, adding 100 mg/L H_2O_2 leads to a large enhancement in the electrical efficiency for all the lamps, which is greatest with the LP lamp. This difference is due to the lower level of light screening introduced by H_2O_2 at 254 nm compared to shorter wavelengths, and it should be noted that at lower levels of H_2O_2, such as might be used in actual water treatment, this difference may not exist because the light screening will not be a large factor. This conclusion is in fact supported by the results of bench-scale reactor testing discussed in Chapter 6. In the case of NDMA, this high level of H_2O_2 did not enhance the performance of any of the lamps and actually decreased the efficiency for MPUV. Thus, these data indicate that the additional oxidant can reduce electrical costs when treating ATZ but it does not appear to convey a particular advantage in the case of NDMA. It should be noted, however, that using UV/H_2O_2 for NDMA treatment does reduce the levels of nitrite formed and leads to more complete mineralization of the compound (Stefan and Bolton 2002).

Table 3.2
Time-based rate constants (min^{-1})

	LP	LP/ H_2O_2	MP	MP/ H_2O_2	P1	P1/ H_2O_2	P2	P2/ H_2O_2
ATZ (D)	0.037	0.25	0.14	0.44	0.29	0.96	0.50	1.3
ATZ (BPI)	0.0086	0.046	0.074	0.223	-	-	-	-
NDMA (D&MWD)	0.16	0.18	0.36	0.31	NC*	NC*	0.88	0.85

*NC: Not calculated due to different pulse rates for runs with and without H_2O_2.
Note: rate constants are not comparable between lamps due to differences in experimental conditions

Comparison of UV Lamps for Direct Photolysis

The cross-laboratory study of ATZ photolysis reveals that comparison of lamps on the basis of k_d' values can reveal which lamps are best suited to a particular application by virtue of their spectral output. When trying to compare lamp efficiencies on such a basis, however, it is important to keep in mind how the UV dose is measured and/or defined, because this will greatly affect the interpretation of the results. As an extreme example, consider what would happen if the UV dose were defined as the energy delivered at 254 nm only. The k_d' values measured for the contaminants might not be so different from those reported here, but when the calculation is performed to see what fraction of the lamp output is actually at 254 nm to determine the electrical requirements for achieving the UV dose defining k_d', the LP lamp would have a huge advantage over the other lamps. In this work, the UV dose has been specified over the range of wavelengths (200-300 nm) most critical to disinfection and to oxidation of the various contaminants investigated. Perhaps more important for immediate purposes is the fact that it is also a range over which the electrical efficiency (conversion of power to photons) of many lamps is specified (Table 1.1).

At various points in this research questions were raised as to whether and how to weight the UV doses. One thought was that introducing a germicidal weighting might make the results more intellectually accessible to water treatment personnel. However, any weighting scheme necessarily biases the definition of UV dose to particular wavelengths as discussed above. In the case of chemical oxidation, the fairest comparison might be to weight the UV dose to the absorption spectrum of the contaminant in question. This would have the benefit of correcting exactly for differences in water transmittance, which are only approximately corrected for by simply using a light screening correction over the 200 to 300 nm range as is done here. Nonetheless, in typical natural waters used as drinking sources with absorbance spectra dominated by nitrate and NOM as in Figure 2.4, the light screening correction should not differ by more than about 10% from that applied in this work. Thus, the k_d' values reported here are likely representative of what will be found in natural waters.

With this background, it becomes possible to compare the lamps using the data in Table 3.1. Considering ATZ first, it is clear that the polychromatic sources lead to much more degradation of ATZ for a given UV dose, and this can be attributed to the higher quantum yield at 220 nm, a wavelength which is not emitted by LP lamps. Comparing the LP and MP lamps at Duke, the difference in rate constants is approximately two-fold, which is similar to the difference in the electrical efficiencies of the two lamp types (Table 1.1). This suggests that the electrical requirements for treating ATZ by direct photolysis with LP and MP should be similar, which is borne out by the results of bench-scale reactor tests described in Chapter 6. Comparison of the MP lamp with the pulsed units shows that the rate constants are somewhat larger for the pulsed lamps, but with the uncertainty in the electrical efficiencies of pulsed lamps (they vary a fair amount depending upon the exact gas composition of the lamp and the applied voltage), it is not really possible at this time to estimate the relative electrical requirements of the different lamps for treating ATZ.

The comparison with the pulsed lamps is also complicated because it is quite possible that the present UV dose measurement is inappropriate with pulsed lamps. The use of a 295 nm long-pass filter allows a good estimate of the 200 to 300 nm irradiance for MP lamps because they do not emit significantly below 200 nm. In contrast, pulsed lamps emit below 200 nm. Thus, the difference between the actinometric photon irradiance (Es cm^{-2} s^{-1}) measured without and with the 295 nm filter represents all photons below 300 nm, and not simply the photons between 200 and 300 nm. When the irradiance is calculated using the 200 to 300 nm PUV spectrum, the conversion of photon irradiance to irradiance (mW cm^{-2}) does not correctly account for the energy of photons below 200 nm. The degree of error introduced by this method is uncertain, but it is clear that actinometry cannot be accurately applied to PUV irradiance measurements unless the emission spectrum and the actinometer quantum yield below 200 nm is known. During the course of this research, it was not possible to acquire this data, and the k_d' values for pulsed lamps must therefore be considered very approximate.

Comparing LP and MP lamps for NDMA treatment leads to a quite different conclusion. The k_d' values for both lamps are very similar, and based on the relative 200-300 nm output efficiency of the lamps, one can estimate that the electrical requirements for NDMA direct photolysis with LP lamps will be approximately half that with MP lamps. Again, this is borne out by the bench-scale reactor results in Chapter 6. The same uncertainties as with ATZ exist when trying to compare the MP lamp to the pulsed units for NDMA. Again, the k_d' values are slightly higher for the pulsed units, but estimating the relative electrical efficiencies is complicated by the variable efficiencies of different pulsed lamps and by the suspected inaccuracies in the UV dose measurement for the pulsed systems.

For three reasons, further research with the pulsed lamps was not pursued. First, the instruments were not readily available at the researchers' labs, which hindered efforts to gather data as required. Second, because the measurement of UV dose for the pulsed systems is not presently well established, any research carried out with these lamps would not lead to definite conclusions. Finally, based on the similar rate constants for direct photolysis of NDMA and ATZ using pulsed and MP lamps and the similarity of the DBP and TOC removal results for the same lamps (Chapter 5), it appeared that no significantly different results could be anticipated between different polychromatic UV sources. This is not to imply that pulsed lamps are the same as MP lamps for water treatment, but simply that in the context of this research there appeared to be no need to pursue their study given the relative difficulty of accessing them and accurately characterizing their performance. Certainly, there is scope for further research into their electrical efficiencies for various applications, and the fact that they do not contain mercury makes them an attractive alternative to traditional lamps whose breakage may lead to public health concerns.

SUMMARY

Photolysis of ATZ and NDMA is very effective with all the lamps investigated. In the case of ATZ, MP and LP lamps are expected to have similar electrical efficiencies while for NDMA LP is anticipated to require less energy. These conclusions are supported by bench-scale reactor results presented in Chapter 6. Allowing for experimental uncertainties in the UV dose measurement for the pulsed lamps, they appeared to perform similar to or slightly better than the MP lamps for both contaminants in terms of electrical requirements. Addition of 100 mg/L H_2O_2 leads to a marked increase in removal rates for ATZ with all the lamps, indicating that adding oxidant will decrease the electrical requirements for ATZ removal. The percent enhancement in ATZ removal rates with H_2O_2 is greatest with the LP lamp, but this difference between lamps may not exist at lower levels of H_2O_2 (e.g., 1 to 10 mg/L) due to reduced light screening. This is borne out by bench-scale reactor testing (Chapter 6). With NDMA, H_2O_2 either had no effect or actually decreased the electrical efficiency of the process. This effect is also observed in bench-scale reactor testing (Chapter 6), for three out of four test scenarios.

Comparison of the mercury vapor lamps with the pulsed lamps is complicated by several factors, including variable energy efficiency of different pulsed lamps and difficulties in accurately quantifying the UV dose for these lamps. For these reasons among others, further research with these lamps was not pursued.

CHAPTER 4
IMPACT OF WATER QUALITY ON DEGRADATION OF CONTAMINANTS

OVERVIEW

To assess the effects of water quality on the UV treatment of contaminants with and without hydrogen peroxide, a two-stage approach was taken. In the first stage, 5 different parameters were varied in a factorial matrix design of experiments to determine the most important factors for subsequent evaluation. In stage 2, only those parameters that were most significant were incorporated into another factorial matrix for more detailed evaluation and statistical analysis.

METHODS

The effect of water quality on the degradation of NDMA, atrazine (ATZ), and MIB, both in the presence and in the absence of hydrogen peroxide was initially investigated in a randomized factorial matrix design presented in Table 4.1, using a collimated beam apparatus equipped with a 1 kW medium pressure Hg lamp. The aimed germicidal UV dose was 1,000 mJ/cm^2, whereas the screened water parameters were pH, total organic carbon (TOC, as natural organic matter), temperature, and alkalinity, as well as dose of H_2O_2.

The experiments were performed in 'synthetic water' buffered at the desired pH using phosphate buffers. The recipes for the 'synthetic water' used in these experiments are described in Chapter 2.

The temperature was monitored during the runs. In the case of 5 and 15 °C experiments, a cooling coil and a recirculating water bath were used as an external mantel to cool the irradiated solution. In order to avoid any stray light coming through reflection/refraction from the coil, the outer wall of the dish was black-painted, and the top of the coil shell covered with black cardboard.

The initial concentration of the contaminants was as follows: NDMA (100 µg/L), ATZ (220 µg/L), MIB (300 ng/L). Over the course of the experiments, the H_2O_2 concentrations remained essentially constant, measured using the method described by Klassen et al. (1994).

STAGE 1 ANALYSIS: WATER QUALITY SCREENING FOR ATRAZINE AND NDMA

Data from the degradation of atrazine and NDMA were statistically analyzed to determine the most significant water quality parameters affecting the degradation of each contaminant. The correlation of rate constants and contaminant destruction with various independent variables was evaluated using Pearson correlation matrix and multiple linear regressions analysis.

Evaluations Based on Contaminant Decay Rates

Pearson Correlation Matrix

This approach was employed to compute a matrix of Pearson correlations among variables, with test of significance of the correlation coefficients obtained by permuting the data in one of the two variables in each pair (Table 4.2).

Table 4.1
Factorial Matrix of Experiments: screening evaluation of water quality parameters on contaminant degradation

Run No.	A H_2O_2, mg/L	B pH, s.u.	C Alkalinity, mg/L as $CaCO_3$	D TOC, mg/L	E Temperature, °C
1	0	6.8	40	2	25
2	15	6.8	40	2	5
3	0	8.2	40	2	5
4	15	8.2	40	2	25
5	0	6.8	240	2	5
6	15	6.8	240	2	25
7	0	8.2	240	2	25
8	15	8.2	240	2	5
9	0	6.8	40	12	5
10	15	6.8	40	12	25
11	0	8.2	40	12	25
12	15	8.2	40	12	5
13	0	6.8	240	12	25
14	15	6.8	240	12	5
15	0	8.2	240	12	5
16	15	8.2	240	12	25
17	*7.5*	*7.9*	***140***	*7*	***15***
18	*7.5*	*7.9*	***140***	*7*	***15***
19	*7.5*	*7.9*	***140***	*7*	***15***
20	*7.5*	*7.9*	***140***	*7*	***15***

Table 4.2
Rectangular Pearson Correlation Matrix

Parameter	H_2O_2	pH	Alk	TOC	Temp
K_{ATZ} (cm^2/J)	-0.255	-0.356	-0.03	-0.241	0.294
% ATZ Removal	-0.297	-0.267	0.016	-0.196	0.346
K_{NDMA} (cm^2/J)	-0.131	-0.025	-0.108	-0.012	-0.296
% NDMA Removal	-0.130	-0.285	0.046	-0.194	-0.234

PEARSON computes and prints matrices of Pearson r correlation coefficients and covariances for all pairs of variables in a list (square matrix option) or for every pair of variables formed by taking one variable from each of two variable lists (rectangular matrix option). Pearson's Correlation Coefficient is usually signified by *r* (rho), and can take on the values from -1.0 to 1.0. Where -1.0 is a perfect negative (inverse) correlation, 0.0 is no correlation, and 1.0 is a perfect positive correlation. SYSTAT 7 software was employed to estimate Pearson Correlation coefficients

Significance of Pearson Correlation Matrix. The correlation table indicates a large scatter of data and no statistical dependence on either the rate constants or on the removal efficiencies of contaminants. However, the following conclusions could be drawn from these data:

1. k_{ATZ} is affected by temperature and pH
2. % ATZ removal is affected by temperature, pH and H_2O_2
3. k_{NDMA} is affected by temperature
4. % NDMA removal is affected by temperature and pH
5. Alkalinity and TOC do not influence the rate constants; however TOC had a slight negative effect on k_{ATZ}

These conclusions are further confirmed by using covariance matrix as summarized in Table 4.3 and discussed below.

Table 4.3
Covariance Matrix

	PEROX	PH	ALK	TOC	TEMP
k_{ATZ} (cm^2/J)	-5.583	-1.156	13.561	-5.571	-13.396
% ATZ Removal	-28.878	-2.445	10.686	-12.701	44.928
k_{NDMA} (cm^2/J)	-0.516	-0.009	-2.941	-0.031	-1.562
% NDMA Removal	-0.864	-0.114	-0.710	-0.546	1.329

Statistical Regression Analysis

Statistical regression of all bench data was conducted using a linear model. The linear regressions are not meant to be predictive but are used to gain insight into significant factors affecting ATZ and NDMA removal. Linear regression was utilized to identify the significance of independent variables.

$$k\ (cm^2/J)) = A + f\ (pH, Alkalinity, H_2O_2, Temperature, TOC)$$

Where k is a rate constant.

Atrazine Data Evaluation: Final Equation in Terms of Actual Factors. The equation developed for the rate constant for atrazine and the corresponding regression statistics was:

$$k_{ATZ}\ (cm^2/J) = 2.95 - 0.0154\ [H_2O_2] - 0.227\ [pH] - 0.00017\ [Alk] - 0.022\ [TOC] + 0.013\ [Temp]$$

The magnitude of the intercept indicates that the independent variables have little or no effect on the rate constant. The relevance of each term was assessed by determining the product of the regression coefficient and the independent variable and normalizing with respect to the intercept. The R square was 0.336 and indicates that 34% of the variability of the predicted ATZ removal can be accounted for by the model. The F statistic term was not very significant (~ 0.41) and indicates the model predictions were not very consistent with the trend of the data. The model indicates that the magnitude of rate constant increased as temperature increased. The effect of alkalinity and TOC was very insignificant. The final regression equation for atrazine can be approximated to:

$$k_{ATZ}\ (cm^2/J) = 2.95 - 0.225\ [pH] + 0.012\ [Temp]$$

This finding is consistent with conclusions drawn from Pearson correlation and co-variance matrix. Thus, it appears that the destruction of ATZ is slightly affected by peroxide, pH and temperature.

Atrazine Regression Statistics	
Multiple R	0.579664821
R Square	**0.336011305**
Standard Error	0.444845878
Observations	17

ANOVA

	df	*SS*	*MS*	*F*	*Significance F*
Regression	5	1.101552068	0.22031	1.113309424	**0.407520303**
Residual	11	2.176766402	0.197888		
Total	16	3.278318471			

	Coefficients	*Standard Error*	*t Stat*	*P-value*	*Lower 95%*	*Upper 95%*
Intercept	2.949946019	1.226512335	2.40515	0.034912917	0.250409205	5.6494828
H2O2, ppm	-0.015366667	0.014828196	-1.03631	**0.322300634**	-0.048003322	0.01727
pH	-0.227332962	0.157451812	-1.44383	**0.176659101**	-0.573882239	0.1192163
Alk, ppm	-0.000167158	0.002162695	-0.07729	**0.939779749**	-0.004927219	0.0045929
TOC, ppm	-0.02185	0.022242294	-0.98236	**0.347034498**	-0.070804984	0.027105
Temp, C	0.0132875	0.011121147	1.194796	**0.25729243**	-0.011189992	0.037765

NDMA Data Evaluation: Final Equation in Terms of Actual Factors. The equation developed for the rate constant for NDMA and the corresponding regression statistics was:

$$k_{NDMA}\ (cm^2/J) = 4.47 - 0.0092[H_2O_2] - 0.016[pH] - 0.0011[Alk] - 0.0012[TOC] - 0.016[Temp]$$

NDMA Regression Statistics	
Multiple R	0.34178379
R Square	**0.11681616**
Observations	17

ANOVA

	df	*SS*	*MS*	*F*	*Significance F*
Regression	5	0.521137495	0.104227	0.290988	**0.908159567**
Residual	11	3.940038976	0.358185		
Total	16	4.461176471			

	Coefficients	*Standard Error*	*t Stat*	*P-value*	*Lower 95%*	*Upper 95%*
Intercept	4.4740103	1.650122806	2.71132	0.020243	0.842112656	8.10590795
H2O2, ppm	-0.0091667	0.01994953	-0.45949	**0.654828**	-0.05307531	0.03474197
pH	-0.0157294	0.211832216	-0.07425	**0.942141**	-0.4819692	0.4505104
Alk, ppm	-0.0011041	0.002909642	-0.37947	**0.711563**	-0.0075082	0.00529996
TOC, ppm	-0.00125	0.029924295	-0.04177	**0.967429**	-0.06711296	0.06461296
Temp, C	-0.015625	0.014962147	-1.0443	**0.318753**	-0.04855648	0.01730648

The relevance of each term was assessed by determining the product of the regression coefficient and the independent variable and normalizing with respect to the intercept. The R square was 0.12 and indicates that 12% of the variability of the predicted NDMA removal can be accounted for by the model. The F statistic term was insignificant (~ 0.908) and indicates the model predictions were not consistent with the trend of the data. Since the variable temperature has the lowest p-value, the most significant parameter is temperature although raw data did not indicate a strong effect of temperature on NDMA destruction. The final regression equation can be approximated to:

$$k_{NDMA}\ (cm^2/J) = 4.45 - 0.015\ [Temp]$$

This finding is consistent with conclusions drawn from Pearson correlation and co-variance matrix. The existing database suggests that the destruction of NDMA has no statistical correlations with variables evaluated. However, we can assume a weak correlation with temperature. More exhaustive experimentation is required to dismiss these conclusions.

Evaluations Based on Contaminant Final Concentration

Statistical Regression Analysis

Statistical regression of all bench data was conducted using a linear model. Linear regression was utilized to identify the significance of independent variables.

$$\%\ ATZ\ or\ NDMA\ Removal = A + f\ (pH, Alkalinity, Peroxide, Temperature, TOC)$$

Atrazine Data Evaluation: Final Equation in Terms of Actual Factors. The equation developed for the percent atrazine removal and the corresponding regression statistics was:

$$\%\ ATZ\ Removal = 103.07 - 0.513\ [H_2O_2] - 4.91\ [pH] + 0.0062\ [Alk] - 0.508\ [TOC] + 0.449\ [Temp]$$

% Atrazine Regression Statistics	
Multiple R	0.564069055
R Square	**0.318173899**
Standard Error	12.91781658
Observations	17

ANOVA

	df	*SS*	*MS*	*F*	*Significance F*
Regression	5	856.5679898	171.3136	1.026629	**0.44816105**
Residual	11	1835.569837	166.87		
Total	16	2692.137826			

	Coefficients	*Standard Error*	*t Stat*	*P-value*	*Lower 95%*	*Upper 95%*
Intercept	103.0700813	35.61651839	2.893884	0.014604	24.6786132	181.4615
H2O2, ppm	-0.51338607	0.430593886	-1.19227	**0.25824**	-1.4611173	0.434345
pH	-4.91032409	4.572220921	-1.07395	**0.30584**	-14.97372	5.153071
Alk, ppm	0.006219113	0.062802191	0.099027	**0.922898**	-0.1320076	0.144446
TOC, ppm	-0.50803243	0.645890829	-0.78656	**0.448153**	-1.9296293	0.913564
Temp, C	0.449283254	0.322945414	1.391205	**0.19166**	-0.2615152	1.160082

The relevance of each term was assessed by determining the product of the regression coefficient and the independent variable and normalizing with respect to the intercept. The R square was 0.32 and indicates that 32% of the variability of the predicted ATZ removal can be accounted for by the model. The F statistic term was insignificant (~0.448) and indicates the model predictions were not consistent with the trend of the data. The P values for the independent variables ranged from 0.19 to 0.92. This again indicates that there is no strong correlation with ATZ removal. P-value indicates pH, peroxide and temperature are the most significant variables. The model indicates that ATZ rejection increased as temperature was increased. The effect by significance and magnitude is temperature, pH, TOC and peroxide, which is indicative of removal of ATZ. The final regression equation becomes:

$$\% \text{ ATZ Removal} = 103.05 - 0.51\,[H_2O_2] - 4.912\,[pH] + 0.45\,[Temp]$$

NDMA Data Evaluation: Final Equation in Terms of Actual Factors. The equation developed for the percent removal of NDMA and the corresponding regression statistics was:

$$\%\text{NDMA Destruction} = 109.73 - 0.099\,[H_2O_2] - 2.329\,[pH] + 0.0062\,[Alk] - 0.222\,[TOC] - 0.133\,[Temp]$$

% NDMA Regression Statistics	
Multiple R	0.4401586
R Square	**0.19373959**
Standard Error	6.20759409
Observations	17

ANOVA

	df	*SS*	*MS*	*F*	*Significance F*
Regression	5	101.8550011	20.371	0.528647	**0.750560844**
Residual	11	423.8764686	38.53422		
Total	16	525.7314697			

	Coefficients	*Standard Error*	*t Stat*	*P-value*	*Lower 95%*
Intercept	109.729477	17.11534514	6.411175	5E-05	72.05883693
H2O2, ppm	-0.0992551	0.206919803	-0.47968	**0.640851**	-0.55468278
pH	-2.3290268	2.197158584	-1.06002	**0.311857**	-7.1649427
Alk, ppm	0.00615799	0.030179288	0.204047	**0.842043**	-0.06026621
TOC, ppm	-0.22285	0.310379705	-0.71799	**0.487732**	-0.90599145
Temp, C	-0.133963	0.155189852	-0.86322	**0.406444**	-0.47553374

The relevance of each term was assessed by determining the product of the regression coefficient and the independent variable and normalizing with respect to the intercept. The R square was 0.19 and indicates that 19% of the variability of the predicted NDMA removal can be accounted for by the model. The F statistic term was insignificant (0.75) and indicates the model predictions were not consistent with the trend of the data as observed in the scatter diagram. P values for independent variables were ranging from 0.31 to 0.84. This again indicates no strong correlation with NDMA removal. P-value indicates pH and temperature are the most significant variables. The model indicates that NDMA rejection increased as temperature was increased.

The effect by significance and magnitude is temperature and UV dose, which is indicative of removal of NDMA. The final regression equation becomes:

$$\%\text{NDMA Destruction} = 109.73 - 2.31\,[\text{pH}] - 0.134\,[\text{Temp}]$$

Stage 1 Conclusions

Based on these statistical evaluations and previous experimental results, it was concluded that the effect of peroxide, pH, and temperature were most significant and should be analyzed in more detail during Stage 2. Various combinations of experimental design were evaluated for Stage 2 analysis and an appropriate randomized matrix was developed as presented in Table 4.4.

STAGE 2 ANALYSIS: DETAILED WATER QUALITY SCREENING – NDMA

Results for NDMA are presented here while a summary of the NDMA and atrazine statistical analyses are presented in Appendix B.

Table 4.4
Stage 2 Matrix for detailed water quality screening

Run No.	H_2O_2, ppm	pH	Temp
1	0.0	7.5	5.0
2	0.0	7.5	15.0
3	5.0	8.5	15.0
4	0.0	6.5	5.0
5	5.0	8.5	25.0
6	5.0	7.5	15.0
7	0.0	8.5	25.0
8	5.0	7.5	5.0
9	5.0	6.5	5.0
10	5.0	7.5	15.0
11	10	8.5	25.0
12	10	7.5	5.0
13	0.0	6.5	25.0
14	10	7.5	15.0
15	0	6.5	15.0
16	0	8.5	5.0
17	10	8.5	5.0
18	5.0	7.5	25.0
19	5.0	6.5	15.0
20	0	7.5	25.0
21	10	8.5	15.0
22	5.0	7.5	15.0
23	5.0	8.5	5.0
24	10	6.5	25.0
25	5.0	7.5	15.0
26	5.0	6.5	25.0
27	10	6.5	5.0
28	0	8.5	15.0
29	10	6.5	15.0
30	5.0	7.5	15.0
31	10	7.5	25.0
32	5.0	7.5	15.0

Note: Run numbers 10, 22, 25, 30, 32 were replicate runs designed to evaluate variability.

Effects of Temperature, pH and Hydrogen Peroxide

Several experiments were conducted at fixed TOC and alkalinity levels, such as 3 and 140 ppm, respectively, whereas, the temperature (3, 13 and 25 °C), pH (6.5, 7.5 and 8.5, buffered solutions), and H_2O_2 concentration (0, 5 and 10 ppm) were varied. These parameters were chosen to mimic as much as possible the surface water characteristics, and their effects on the pseudo-first order rate constant of NDMA photo-induced degradation are given in Figure 4.1.

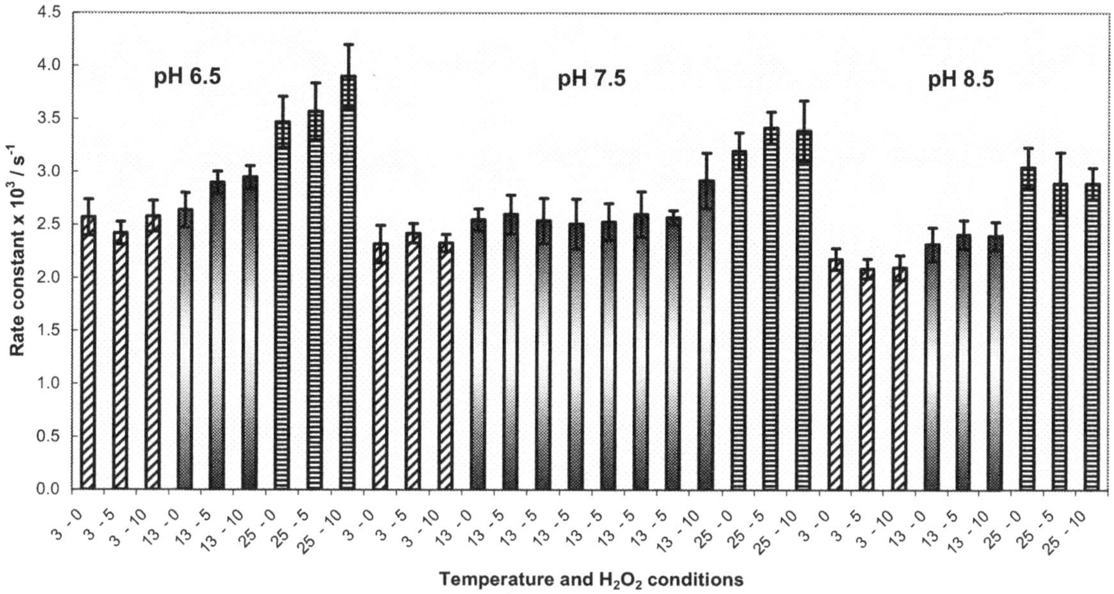

Figure 4.1 Rate constants of NDMA UV light-induced degradation under various pH, temperature and H_2O_2 conditions. The first number in the x-axis 'scale' is the temperature and the second number is the H_2O_2 concentration in ppm

Irrespective of pH and H_2O_2 concentration, the rate constants increase as the temperature increases. Table 4.5 shows the activation energies as calculated from the Arrhenius plots, as a function of pH and H_2O_2 concentration. The activation energy depends slightly on the H_2O_2 concentration and on the pH. In the photolysis of NDMA, such activation energies are probably required for the hydrolysis of the NDMA excited state. An about 40% increase in the rate constant is calculated when the temperature increases from 3 to 25 °C, at any of the experimental pH values.

Table 4.5
Arrhenius Activation Energy for the Photolysis of NDMA

	E_a/ kJ mol^{-1}		
H_2O_2 / ppm	pH 6.5	pH 7.5	pH 8.5
0	9.5	10.1	10.5
5	12.1	10.9	10.1
10	12.9	11.6	10.0
Average	**11.5**	**10.9**	**10.2**
Standard deviation	**1.8**	**0.8**	**0.3**

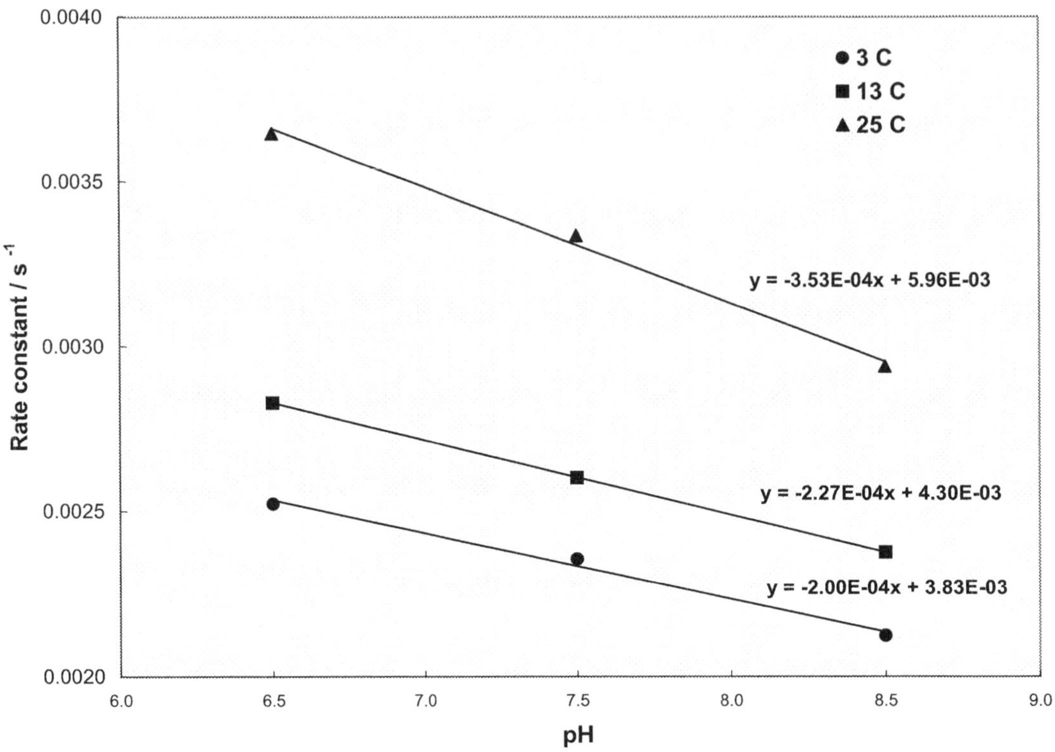

Figure 4.2 Effect of pH on the pseudo-first order rate constant of NDMA photo-degradation

As a general trend, at a given temperature, and provided a relatively insignificant effect of H_2O_2 concentration, the rate constant decreased as pH increased, by about ~10-12% per pH unit. This effect is illustrated in Figure 4.2. Again, for a narrow pH range such as 7.7-8.2, as usually encountered in the environmental water sources, one can say that there is no significant impact of pH on the photolysis rate of NDMA.

When irradiated over the 200-300 nm band, hydrogen peroxide photolyzes generating the powerful oxidative •OH radicals with a quantum yield of 1.0. In the presence of H_2O_2, NDMA undergoes both direct photolysis and •OH radical-induced degradation, with a rate constant of 3.3×10^8 M^{-1} s^{-1} (Wink et al., 1991). Providing the large absorption coefficients and quantum yield of NDMA in the 200-300 nm range, and the competition for the light and •OH radicals of the water background (1.9–2.7 $\times$ 10^4 mg C L^{-1} s^{-1}), large concentrations of H_2O_2 would be required to overcome such competitions. Therefore, at the operational H_2O_2 concentrations in the water treatment, which are similar to those used in this study, a relatively low impact of •OH radical-driven processes on the NDMA decay is expected.

According to the above results, the only significant H_2O_2 concentration effect on the rate constant was observed at pH 6.5, 25 °C, where an increase of ~25% was induced by the addition of 10 ppm H_2O_2. This may be explained by a lower concentration of bicarbonate ions at pH 6.5 as compared to those at pH 7.5 and 8.5, and therefore, a lower competition for the •OH radicals due to these ions. As expected, a temperature effect on the rate constant of NDMA decay in the presence of H_2O_2 was observed.

Statistical Evaluation of Water Quality Parameters Effect on NDMA Photolysis

The experimental data provided in Figure 4.1 was also evaluated by multiple regression, covariance coefficients, and Pearson correlation matrix, based on the same variables as above: pH, temperature, and H_2O_2 concentration. Table 4.6 gives the statistical parameters, followed by the equation describing the predicted rate constant of NDMA photodegradation based on the multiple linear regression analysis. The multiple linear regression constants are: multiple R = 0.9324; R^2 = 0.8695; adjusted R^2 = 0.8555; standard error = 0.000249; no. of observations = 32.

Table 4.6
Statistical Parameters for NDMA photo-induced degradation

H_2O_2	pH	T
Pearson Correlation Matrix		
0.139875	-0.40085	0.830184
Covariance Matrix		
0.000265	-0.00026	0.003911

$$k_{NDMA}\ (\times 10^5) = 164.77 + 8.69\ [H_2O_2]_0 - 8.74\ \text{pH} + 2.20\ \text{T}$$

The coefficients obtained from this evaluation indicate a positive effect of peroxide and temperature and an inverse effect on pH. The multiple R is 0.93 and indicates that the models can account for 93% of the variability of the predicted destruction. The F statistic term was significant and indicates the model predictions are consistent with the trend of the data. The model indicates that NDMA removal increased as temperature and peroxide increased. The major effect by significance and magnitude is temperature and peroxide, which are indicative of a radical path mechanism.

IMPACT OF WATER QUALITY ON MIB DEGRADATION

The impact of water quality on the UV advanced oxidation destruction of MIB was run in separate experiments at Duke and evaluated at UCLA independent of the NDMA and atrazine results.

Experimental Results

Table 4.7 presents experimental results for laboratory scale studies on the degradation of MIB. In particular, MIB concentration for initial and treated samples is summarized. Initial samples were sampled in duplicate from the same batch water and represent spiked water before irradiation was applied. Two separate aliquots were collected from the batch water and exposed to UV light. One sample was exposed to a UV dose of 400 mJ/cm^2, while the other was exposed to a UV dose of 900 mJ/cm^2. Treated samples were also sampled in duplicate from the same batch after exposure to UV light. Thus, influent samples were compared to effluent samples treated at two separate UV conditions, but at the same temperature, pH, and H_2O_2 dose.

Percent removal was computed from the initial and treated MIB concentration determined. The two duplicate initial sample concentration values were averaged together, whereas the treated sample values were not. The following equation describes the computation for percent removal:

Percent removal = (avg. initial conc. – treated conc.) / (avg. initial conc.)

A four factor (pH, temperature, UV dose and H_2O_2) experimental design was developed. Each experimental factor was set at two levels, a low and a high value as follows: 1) pH 6.5 and 8.5; 2) Temperature 10 and 25°C; 3) UV dose at 450 and 900 mJ/cm^2; 4) Hydrogen peroxide dose at 5 and 15 mg/L. For the ANOVA modeling, high values received a value of 1, while the low values received a value of 0. Thus, for the statistical analysis, percent removal of MIB was the dependent variable. Temperature, pH, UV dose and H_2O_2 dose alone and various combinations were the dependent variables.

Figure 4.3 plots each percent removal according to each treatment received. The plot roughly describes the percent removal for each treatment, but does not account for interacting effects, nor does it include error bars. Thus, definitive statements cannot be made based on the plot. However, it is interesting to notice that treatments 1a and 1b (high H_2O_2, UV, temperature, and pH) and 2a and 2b (high H_2O_2, UV and temperature) correspond to the highest percent removal of MIB. The plot illustrates how the quadruplicate runs in experiments 4, 5, 7 and 15 have apparent differences in percent removal of MIB.

Quality Assurance

For all the experiments, influent water was the same for two different effluent samples, meaning that there were eight influent samples and sixteen effluent samples analyzed. Therefore, MIB influent concentration is the same for the following samples: one and five, two and six, three and seven, four and eight, nine and thirteen, ten and fourteen, eleven and fifteen, twelve and sixteen. Of the eight influent samples, five samples were run in duplicate: 1) one and five, 2) three and seven, 3) nine and thirteen, 4) eleven and fifteen, 5) twelve and sixteen. The average difference of the samples run in duplicate is 75 ng/L. The standard deviation for all the duplicate samples is 70 ng/L. Two of the influent samples presented in Table 4.7 show larger values than the rest of the samples: 1) 9a and 13a, and 2) 11a and 15a. Since influent samples 11a and 15a have low treatment conditions, the ANOVA evaluation of determining the optimum parameters should not be affected.

Table 4.7
Bench Scale Experiment Results - MIB

No Treatment Applied			*Treatment Applied*						
Sample No. (Spiked initial)	Observed Initial MIB (ng/L)	Averaged initial MIB (ng/L)	Sample No. (Treated)	Observed Treated MIB (ng/L)	H2O2 (mg/L)	UV (mJ/cm^2)	Temp (deg. C)	pH	Removal Efficiency of MIB
1a	205	212	1a	53	15	900	25	8.5	0.75
1b	219		1c	59	15	900	25	8.5	0.72
2a		233	2a	46	15	900	25	6.5	0.80
2b	233		2c	45	15	900	25	6.5	0.81
3a	236	223	3a	58	15	900	10	8.5	0.74
3b	210		3c	83	15	900	10	8.5	0.63
4a		190	4a	109	15	900	10	6.5	0.43
4b	190		4c	137	15	900	10	6.5	0.28
5a	205	212	5a	100	15	450	25	8.5	0.53
5b	219		5c	130	15	450	25	8.5	0.39
6a		233	6a	96	15	450	25	6.5	0.59
6b	233		6c	96	15	450	25	6.5	0.59
7a	236	223	7a	78	15	450	10	8.5	0.65
7b	210		7c	126	15	450	10	8.5	0.43
8a		190	8a	62	15	450	10	6.5	0.67
8b	190		8c	64	15	450	10	6.5	0.66
9a	311	258	9a	121	5	900	25	8.5	0.53
9b	205		9c	112	5	900	25	8.5	0.57
10a		240	10a	135	5	900	25	6.5	0.44
10b	240		10c	112	5	900	25	6.5	0.53
11a	437	347	11a	120	5	900	10	8.5	0.65
11b	256		11c	138	5	900	10	8.5	0.60
12a	243	219	12a	152	5	900	10	6.5	0.31
12b	194		12c	127	5	900	10	6.5	0.42
13a	311	258	13a	201	5	450	25	8.5	0.22
13b	205		13c	203	5	450	25	8.5	0.21
14a		240	14a	173	5	450	25	6.5	0.28
14b	240		14c	140	5	450	25	6.5	0.42
15a	437	347	15a	168	5	450	10	8.5	0.52
15b	256		15c	208	5	450	10	8.5	0.40
16a	243	219	16a	113	5	450	10	6.5	0.48
16b	194		16c	117	5	450	10	6.5	0.47

*Computation for removal efficiency: (avg. initial conc. - treated conc.) / avg. initial conc.

*Treatment (H202, UV, Temp, pH) are the independent variables and per cent removal is the dependent variable for the ANOVA analysis

Statistical Analysis

The duplicate values for MIB initial concentration were averaged in the percent removal calculation as the same sample was sampled in duplicate. The treated values were treated as unique to have greater power for ANOVA and to account for within sample variation. Thus, the entire sample set for effluent was used for the analysis, and each replicate was independently treated from the other. To verify the consistency of the duplicate results, a paired t-test was performed in Excel. If the resulting p-value was less than the alpha-value of 0.05, then the null hypothesis stating that there is no significant difference between each duplicate set must be rejected. However, the two-tailed p-value computed for this data was 0.26, indicating that there is no statistical difference between duplicates. Therefore, the t-test shows that there is no difference between effluent replicates and verifies that each datum can be evaluated separately.

Assumptions for the t-test test are that the data consist of a sample of independent measurements that are normally distributed with unknown mean and fixed, finite variance. Visual inspection of diagnostic plots verifies fixed finite variance and that the sample did not deviate significantly from normality.

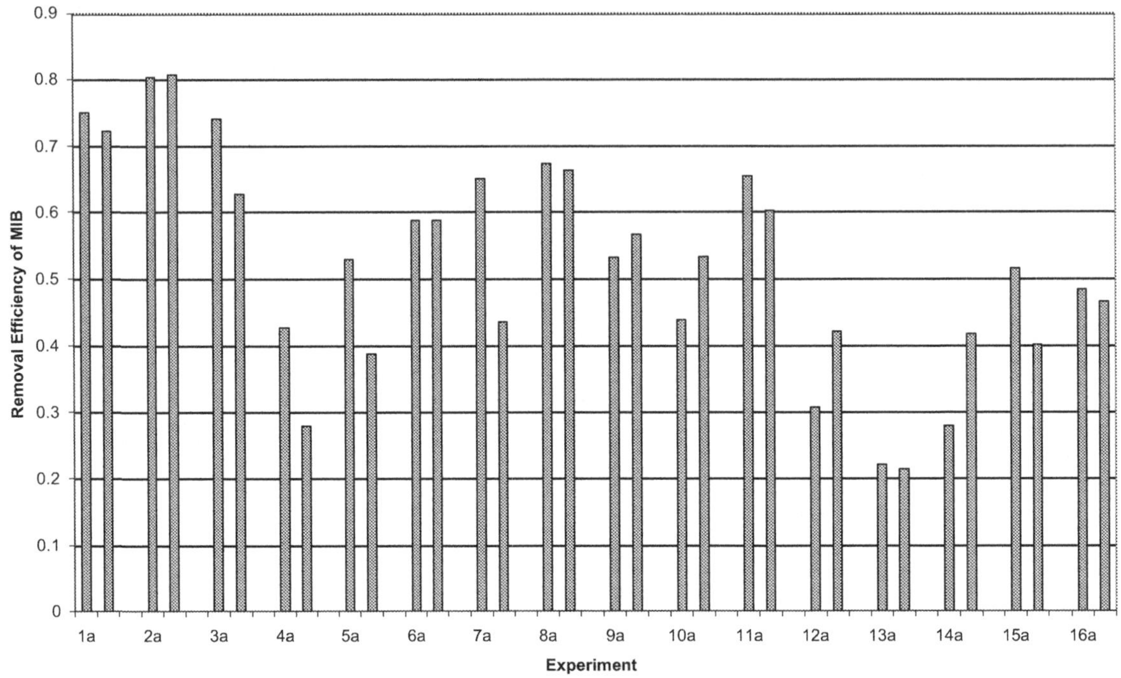

Figure 4.3 Percent Removal (MIB) per Experiment from Raw Data

A four factor (pH, temperature, UV dose, and H_2O_2) experimental design was developed. Each experimental factor was set at two levels, a low and high value as follows: 1) pH: 6.5 and 8.5; 2) Temperature: 10 and 25° C; 3) UV dose: 450 and 900 mJ/cm^2; 4) Hydrogen peroxide dose: 5 and 15 μg/L. For the ANOVA modeling, high values received a value of 1, while the low values received a value of 0. Thus, for the statistical analysis, percent removal of MIB was the dependent variable. Temperature, pH, UV dose, and hydrogen peroxide (H_2O_2) dose alone and in various combinations were the independent variables.

An ANOVA evaluation was performed using the S-Plus statistical package in order to determine which factors alone or in combination were significant in the treatment of MIB. Since there were four individual treatment variables, there were 16 potential treatment combinations for modeling. The system was coded for the independent variables so that the low values for each condition received a value of 0, whereas the high values received a value of 1. The significance of this process should be apparent in the following discussion. To perform an ANOVA, the assumptions are that the replicates are independent measurements and that the residual errors are independent and normally distributed with fixed variance. The replicates are indeed independent since each measurement represents MIB concentration from a different bottle. In order to verify the independence and normality of the residuals, diagnostic plots for the fit of the data to the model were scrutinized. In addition, a Kolmogorov-Smirnoff goodness of fit test of normality of the residual error distribution was performed, indicating that the residuals did not deviate significantly from normality (p-value=0.5).

Results

An ANOVA test to determine the controlling parameters and interacting parameters was completed for the data set. Main effects and interaction plots were analyzed for each treatment factor. The main effects plots are shown as box and whisker plots in Figure 4.4. Since interactions were found, interpretation of the main effects in Figure 4.4 is only for illustrative purposes. Figure 4.5 graphically shows interactions between treatment factors. The final interpretation of the model is represented in Figure 4.6.

The main effects in Figure 4.4 roughly illustrate how each treatment factor individually affects the percent removal of MIB, but more importantly, illustrates the normality of the data. Gross skewing for percent removal is not apparent. Minor skewing illustrated in the plots of Figure 4.4 is expected due to small sample size. In all cases, the higher treatment condition seems to have an increased percent removal in comparison to the lower treatment conditions, but overlapping boxes indicate the increase in removal is not statistically significant. Of all four main treatment factors, increased H_2O_2 seems to have the greatest effect on MIB removal. However, the ANOVA interpretation ultimately explains what factors alone or in combination can optimize MIB removal.

Significance of experimental factors was quantified by generating the ANOVA results, outlined in Table 4.8. To determine significant interactions, a full model with all possible treatment combinations was fitted to the data, computing an F-statistic with its corresponding p value. From this model, all non-significant interactions were removed. The key result from the ANOVA table is that the p-values for the F statistics corresponding to statistically significant experimental factors and their interactions are less than the alpha level of 0.05, thus indicating that such treatment significantly reduced MIB. For these data, the factors with a p value less than 0.05 included the following treatments set at the high condition: 1) pH:Temperature:UV; 2) Temperature:H_2O_2; 3) Temperature:UV; 4) pH:UV; 5) pH:Temperature; 6) H_2O_2; and 7) UV. Yet, since individual components of a significant combination must be modeled, pH and temperature were included as coefficients as well for the linear regression model despite having p-values of 0.36 and 0.93 respectively. However, treatment consisting of UV and H_2O_2 was not significant, having a p value of 0.32, which is why UV: H_2O_2 is not included as a separate coefficient in the linear regression model presented below in Table 4.9.

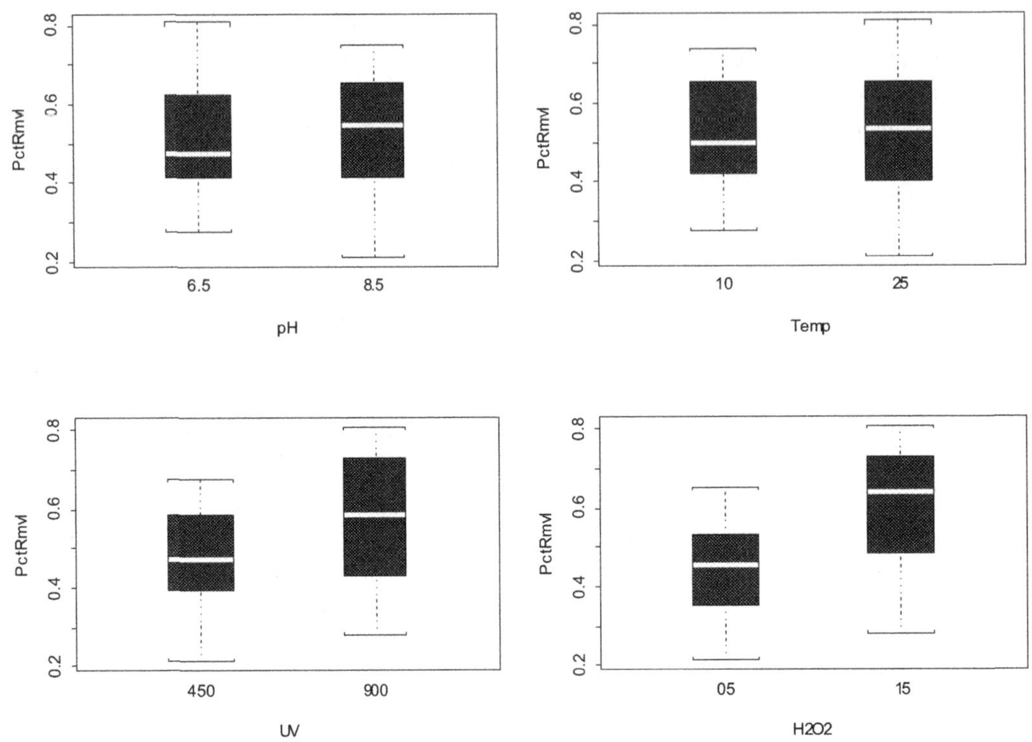

Figure 4.4 Main Effects Plots for Each Experimental Factor at High and Low Levels Versus Per Cent Removal

Note: The horizontal line inside the box represents the median, the upper edge is quartile 3 (Q_3), and the lower edge is quartile 1 (Q_1). The distance between quartile 1 and 3 is called the interquartile range (IQR). The vertical lines above and below the box are the whiskers, and the horizontal lines above the whiskers are the fences. The upper fence is at the largest value, or at (Q_3 + 1.5 (IQR)), whichever is lower. Likewise, the lower fence is at (Q_1 – 1.5(IQR)) or the smallest value, whichever is higher.

Table 4.8
Analysis of Variance statistical data

Experimental Factors	Df	Sum of Sq	Mean Sq	F Value	P Value
pH	1	0.00	0.00	0.88	0.361
Temp	1	0.00	0.00	0.01	0.933
UV	**1**	**0.09**	**0.09**	**17.91**	**0.001**
H2O2	**1**	**0.22**	**0.22**	**43.09**	**0.000**
pH:Temp	**1**	**0.07**	**0.07**	**13.02**	**0.002**
pH:UV	**1**	**0.12**	**0.12**	**24.69**	**0.000**
pH:H2O2	1	0.00	0.00	0.77	0.393
Temp:UV	**1**	**0.15**	**0.15**	**29.11**	**0.000**
Temp:H2O2	**1**	**0.05**	**0.05**	**10.87**	**0.005**
UV:H2O2	1	0.01	0.01	1.06	0.318
pH:Temp:UV	**1**	**0.03**	**0.03**	**5.84**	**0.028**
pH:Temp:H2O2	1	0.00	0.00	0.19	0.670
pH:UV:H2O2	1	0.00	0.00	0.04	0.835
Temp:UV:H2O2	1	0.01	0.01	1.67	0.215
pH:Temp:UV:H2O2	1	0.01	0.01	2.35	0.145
Residuals	16	0.080	0.005		

Interaction plots, shown in Figure 4.5, graphically illustrate interactions between treatment factors. Three interaction plots at pH of 6.5 and three interaction plots at pH 8.5 are presented. The x-axis corresponds to treatment, and the y-axis corresponds to percent removal of MIB. The lines plotted correspond to a second treatment factor at two different conditions. Parallel lines mean there is no interaction between the two factors diagrammed. At pH 6.5, the lines are roughly parallel for the plot representing H_2O_2 and UV, meaning that there is no interaction between the two treatment factors. However, there is an interaction between UV and temperature since the lines cross, and an interaction between H_2O_2 and temperature is indicated, since the lines are not quite parallel, and almost intersecting. The interactions are similar at pH of 8.5. The interaction, however for UV and temperature is not so obvious since the lines do not actually cross in the plot. For H_2O_2 and temperature, the lines almost cross. Finally, the lines are obviously more parallel for UV and H_2O_2 at pH 8.5 than they are for pH 6.5. Therefore, the consistency of the interactions shown plotted in Figure 4.4 at two different pHs supports the results generated by ANOVA.

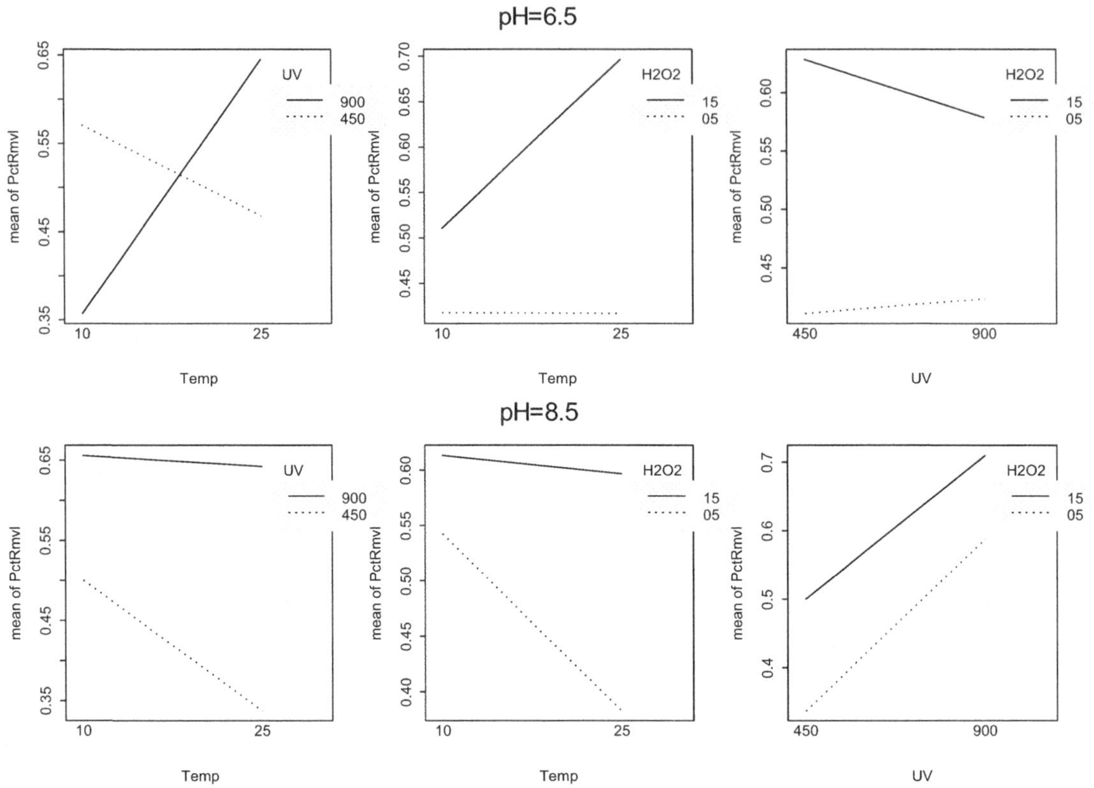

Figure 4.5 Interaction Plots for Temperature, H_20_2, and UV at pH 6.5 and 8.5

Linear Regression Model

Table 4.9 outlines the linear regression model formulation from the nine significant variables generated in the ANOVA analysis. Contents of Table 4.9 include the model coefficient, standard error, t-value and the p-value for each predictor included in the model. The estimated average removal efficiency for each treatment condition is determined from applying the model using the appropriate parameters to the experimental conditions under study. The model describing average population percent removal for all the significant variables specified in the ANOVA analysis is:

% removal = a + b_1*x1 + b_2*x2 + b_3*x3 + b_4*x4 + b_5*(x1:x2) + b_6*(x1:x3) + b_7*(x2:x3) + b_8*(x2:x4) + b_9*(x1:x2:x3) + residual error

Where a is the intercept (the average per cent removal when all four factors are set at the low level), and b_1-b_9 are the coefficients determined by the linear regression model via the least squares method. In the model computed, x_1 represents pH, x_2 represents temperature, x_3 represents UV and x_4 represents H_2O_2. Depending if the treatment is at a low or high condition, x_1 through x_4 would be 0 or 1. Table 4.9 presents the coefficient for the linear regression model for the treatment combination at high conditions: pH:Temp:UV. The resulting linear regression model must include all components of any given combination. Thus, the model to describe MIB removal with pH:temp:UV would be:

Estimated population % removal = 0.53 + (-0.07)*1 + (-0.19)*1 + (-0.21)*1 + (-0.08)*1 + (-0.06)*1 + (0.37)*1 + (0.39)*1 + (0.17)*1 + (-0.24)*1 = 0.56

It is interesting to note that while several of the individual factors (pH, temp, UV, H_2O_2) and combinations of factors (pH:temp, pH:temp:UV) have negative coefficients, however the positive of effects of pH, temp:UV, and temp:H_2O_2 increase the estimated % removal to 0.56.

Table 4.9
Linear Regression Model Formulation for MIB Removal

Variable	Value	Std. Error	T-value	P-value
Intercept	0.53	0.04	13.37	0.000
pH	-0.07	0.05	-1.42	0.171
Temp	-0.19	0.06	-3.31	0.003
UV	-0.21	0.05	-4.27	0.000
H202	0.08	0.04	2.31	0.000
pH:Temp	-0.06	0.07	-0.84	0.410
pH:UV	0.37	0.07	5.21	0.000
Temp:UV	0.39	0.07	5.51	0.000
Temp: H202	0.17	0.05	3.29	0.003
pH:Temp:UV	-0.24	0.10	-2.41	0.025

Comparison of High and Low Treatment Doses

The next step in this analysis was to determine how the experimental factors set at the highest doses compared with the lowest treatment dose. Figure 4.6 illustrates the 95% confidence limit error bars computed from the standard error determined for each parameter and combination of parameters for the linear model. Any sample having a treatment combination with error bars overlapping with the lowest treatment dose is not significantly different from samples receiving no treatment. The top bar in the plot represents the predicted percent removal for treatment conditions at the low values. For this data, the 95% confidence interval for MIB removal with the lowest treatment conditions is a removal efficiency of 0.53± 0.08 since x_1 through x_4 would be 0. Therefore, when all treatment conditions are applied at the low condition, the percent removal equals a, the intercept from the linear regression model.

Figure 4.6 presents all treatment combinations, but the percent removal is computed only from the coefficients determined to be significant by ANOVA. For instance, when temperature, UV, and H_2O_2 are at the high condition, the model would only include the coefficients presented in Table 4.8. There is not a coefficient representing temperature, UV, and H_20_2 all together, since that combination is not significant. However, the significant components of the three factors could be modeled. Thus, the model describing the estimated percent removal of any population of water samples with similar conditions applied would be:

% removal = a + b_2*Temp + b_3*UV + b_4* H_20_2 + b_7*(Temp:UV) + b_8*(Temp:H_20_2) + residual error

In numeric form, using the coefficients generated from the linear regression model shown in Table 4.9, the computation for percent removal would be:

% removal = 0.53 + (-0.18)*1 + (-0.21)*1 + (0.08)*1 + (0.39)*1 + (0.16)*1 = 0.76

Judging from Figure 4.6, the three most successful treatments of MIB are the following combinations: 1) Temperature, UV, and $H_2 0_2$; 2) pH, temperature, UV and $H_2 0_2$; and 3) pH, UV and $H_2 0_2$. It is interesting to note the removal efficiencies for 1) Temperature, UV and $H_2 0_2$, and 2) pH, temperature, UV and $H_2 0_2$ are roughly equivalent: 0.75±0.08, surprisingly indicating a basic pH may not be an essential factor for treatment of MIB when a higher temperature is applied. Additionally, the removal efficiency with pH, UV, and $H_2 0_2$ at the higher conditions is approximately 0.6±0.08, lower than the removal efficiency for temperature, UV and $H_2 0_2$. However, the error bars do overlap, indicating the treatment using the combinations of temperature, UV and $H_2 0_2$; and pH, UV and $H_2 0_2$ are not statistically different from each other. In any case, the results from the linear regression model clearly indicate treatment of MIB using $UV/H_2 0_2$ is optimized under basic pH or high temperature. Therefore, the results agree with the initial hypothesis made: basic pH or high temperature would enhance the degradation rate of MIB with $UV/H_2 0_2$ since it would increase the production of the hydroxy radical. Yet, statistical analysis does not support a further claim made in the hypothesis: high temperature should remove significantly more MIB in comparison to a basic pH.

Figure 4.6 also illustrates the factors having lower removal efficiencies than all four treatment factors set at the lowest conditions. Interestingly, temperature alone to 25 °C, UV alone at 900 nm, and pH and temperature at 8.5 and 25 °C, were predicted to have removal efficiencies of 0.35±0.08, 0.33±0.08, and 0.22±0.08, thus significantly lower than all parameters at low treatment values.

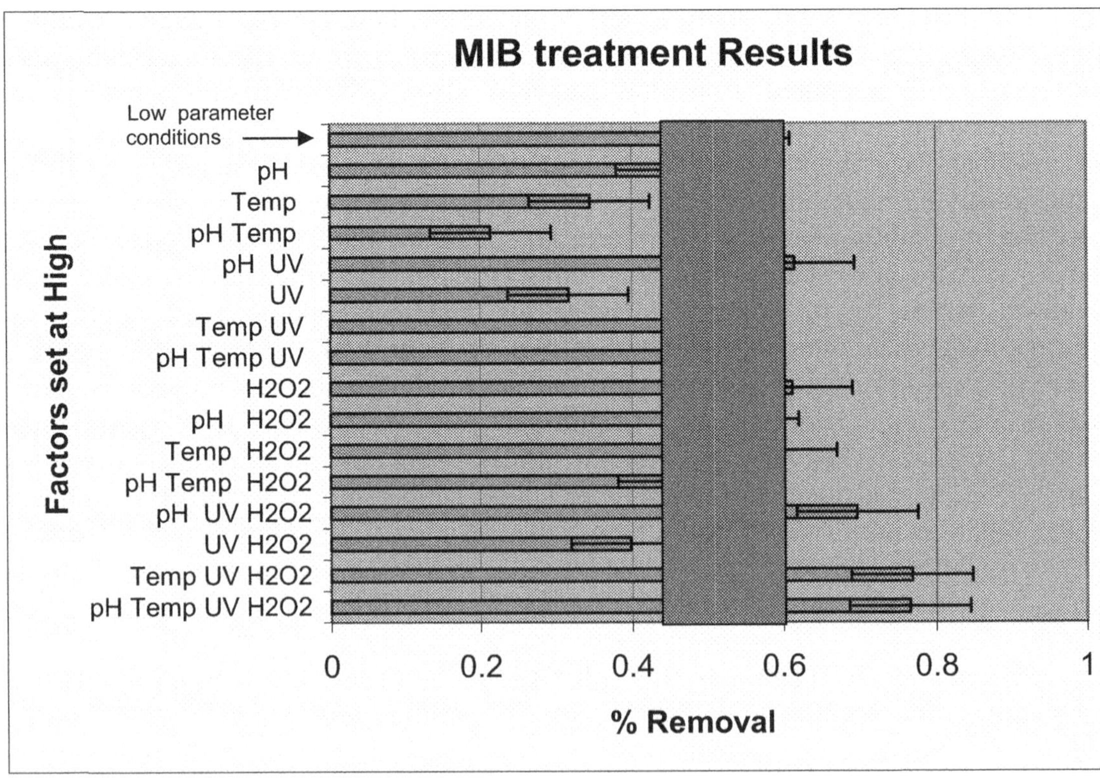

Figure 4.6 Plot of Modeled Percent Removal for MIB. Note: Top bar corresponds to computed per cent removal with the lowest treatment conditions. Remaining bars correspond to specific factors and combinations of factors set at high. Shaded region corresponds to error bar region from low treatment condition.

Figure 4.7 is a scatter plot depicting observed versus predicted MIB removal values. Observed values are shown on the x-axis and predicted values are shown on the y-axis. Since replicate treated samples were modeled, one single predicted value corresponds to two observed values. An R^2 value of 87% describes the degree of correlation between the observed and modeled values and indicates that the model has a linear relationship with the actual lab data. Therefore, the model adequately predicts percent removal of MIB.

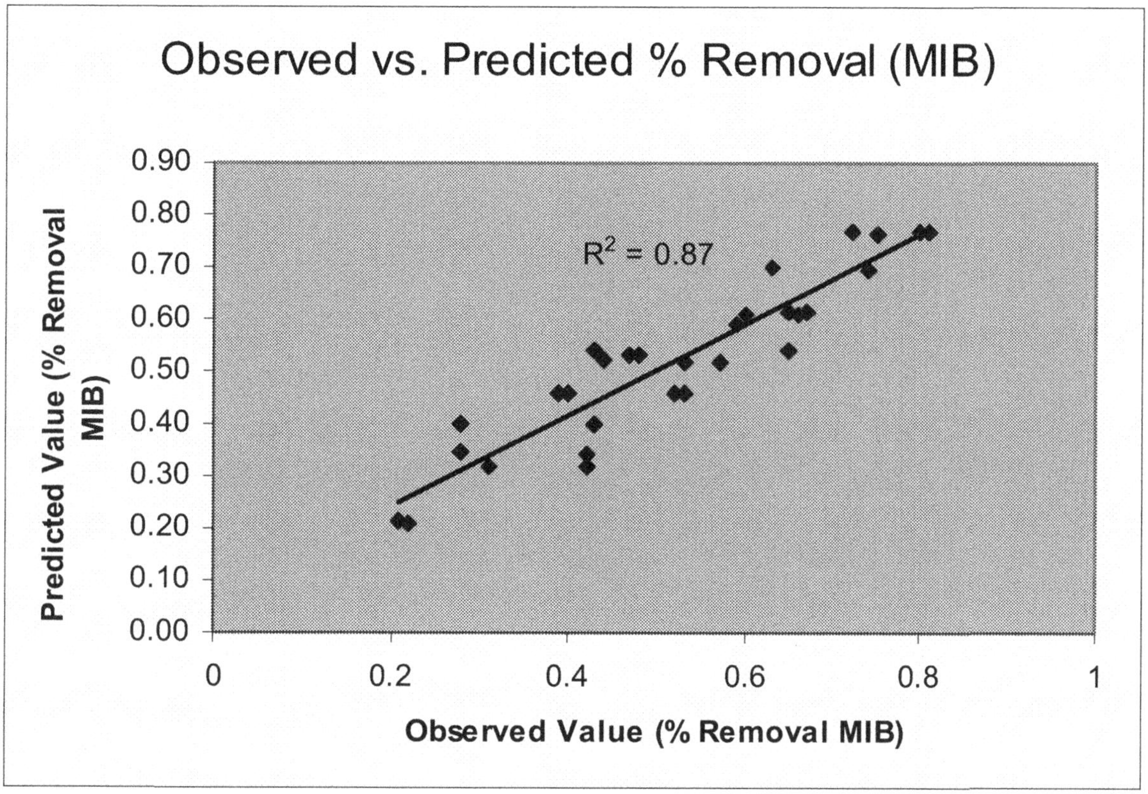

Figure 4.7 Observed vs. Predicted Correlation Plot for percent Removal MIB

Conclusion

In conclusion, ANOVA analysis followed by the formulation of a linear regression model of the data showed UV/H_2O_2 had an enhanced effect on the removal of MIB when combined with pH or temperature. The effect of temperature was stronger than the effect of pH, but statistical analysis showed that the effects are not significantly different from each other.

CHAPTER 5
IMPACT OF UV PROCESSES ON FORMATION OF DBPS

INTRODUCTION

The impact of different UV technologies combined with H_2O_2 on disinfection byproduct (DBP) formation was investigated using a broad range of UV doses. Four classes of DBPs (THMs, HAAs, aldehydes and carboxylic acids) were examined. This enabled the determination of effects resulting from the action of a UV-based Advanced Oxidation Technology (AOT; UV/H_2O_2 oxidation) on natural organic matter (aldehydes, carboxylic acids) as well as effects on the ultimate formation of chlorinated DBPs (THMs and HAAs) from secondary chlorination.

Much of the experimentation was performed using synthetic water, described in Chapter 2, in order to have more control of the many variables that would affect DBP formation. Additional testing was performed with synthetic water that had been augmented with nominally 100 μg/L bromide, and with filtered natural water from the Grand River, Kitchener ON Canada. Experiments with filtered natural water were performed at both bench-scale and pilot-scale. The data presented in this Chapter are augmented by several appendices (C, D, E, F, G) providing further information on the DBP testing performed including (C) Effect of high dose MP UV irradiation with and without peroxide; (D) Effect of temperature; (E) Effect of bicarbonate; (F) Effect analysis of factorial matrix; and (G) Example of raw data collected.

In general, for doses of less than 500 mJ/cm^2, UV/H_2O_2 oxidation did not significantly affect THM and HAA formation in subsequent chlorination processes, however higher UV doses resulted in much lower ultimate concentrations of THMs and HAAs (approximately 50% lower). UV/H_2O_2 oxidation also resulted in the formation of aldehydes and carboxylic acids at UV doses above 500 mJ/cm^2, compounds that are known to adversely effect drinking water biostability. Compared to UV irradiation alone at similar doses, UV/H_2O_2 oxidation had a more significant impact on DBP formation.

EFFECTS OF DIFFERENT UV TECHNOLOGIES ON DBP FORMATION – BENCH TESTING WITH SYNTHETIC WATER

Samples of synthetic water (350 mL) prepared in the Duke, BPI and MWD laboratories (Table 2.1) were placed in a 125x65 mm crystallization dish (sample depth approximately 3 cm) and irradiated with either MPUV, LPUV or P-UV with or without approximately 100 ppm H_2O_2. The Duke, BPI and MWD laboratories were designated as A, B, C, respectively, in this chapter. The high concentration of peroxide was chosen to ensure advanced oxidation conditions for these experiments. Initial H_2O_2 concentrations were verified by the colorimetric method described in Section 2. The UV doses were also determined as described in Section 2, and at approximate doses of 0, 40, 140, 500, 1000, and 5000 mJ/cm^2 the irradiations were stopped and the samples were prepared for DBP analysis. The lower doses represented those that may be encountered during disinfection, while the higher doses represented those used in oxidative applications. When H_2O_2 was present in the samples, 0.2 mg/L catalase was added immediately after irradiation to quench the residual H_2O_2. Experiments performed at the University of Waterloo (Appendix A) showed that this level of catalase does not contribute significantly to the level of DBPs, nor does it noticeably increase the TOC concentration. Irradiations were generally performed in duplicate except for the 40 and 1000 mJ/cm^2 doses, which were performed in triplicate. Samples were refrigerated, and the day following irradiation, they were shipped on ice in coolers to the University of Waterloo for uniform formation condition (UFC) testing and DBP analyses.

DBP analyses and UFC testing were initiated upon receipt of the samples at the UW laboratory. Aldehyde concentrations were determined by GC-ECD analysis of extracted PFBHA derivatives. Carboxylic acid concentrations were determined by ion chromatographic analysis of chloroform-preserved samples. THM and HAA concentrations were determined following UFC testing. UFC pretests were employed to determine the chlorine dose that would result in a 24 hour residual of 1.0 ± 0.4 mg/L as Cl_2. For most of the samples described in this report, a chlorine dose of 4.5 mg/L as Cl_2 consistently resulted in 24 hour residual of 1.0 ± 0.4 mg/L as Cl_2. Residual chlorine was quenched with sodium thiosulfate immediately at the conclusion of the 24 hour reaction period, following which THM and HAA analyses were performed using standard methods. Travel standards and travel blanks for each class of DBP were included with each batch of samples. Laboratory standards and blanks were also prepared at the time of analysis of each set of samples.

The MP UV experiments with synthetic water were performed at both the Duke and BPI laboratories, designated as laboratories A and B, respectively, providing a measure of the comparability of the results obtained from the different laboratories. The P-UV experiments were performed at both the Duke and MWD laboratories (laboratories A and C, respectively). Different P-UV systems were used in laboratories A and C as discussed in Chapter 2.

In general, neither LP UV, MP UV nor P-UV had a significant effect on the formation of DBPs at doses of 500 mJ/cm^2 or less (UV disinfection doses). At or above doses of 1000 mJ/cm^2, however, LP UV, MP UV and P-UV caused changes in the NOM of the synthetic water such that DBP concentrations were affected to similar degrees. Also, in the presence of approximately 100 mg/L H_2O_2 (AOP conditions), LP UV/H_2O_2, MP UV/H_2O_2 and P-UV/H_2O_2 resulted in significant changes to the NOM in the synthetic water such that DBP concentrations were affected.

The following sections contain summaries of the results obtained for these experiments. Within each section, the first two figures (a/b) are different depictions of the DBP concentrations obtained from the UV irradiation of synthetic water (± UFC testing). One figure provides the reader with the raw data. It has been plotted on a split linear scale to illustrate direct correlations between UV dose and DBP concentration (if any) and the relative amount of scatter in the data for both 'low' UV doses (<500 mJ/cm^2) and 'high' UV doses (>1000 mJ/cm^2). The second of these two figures provides averages of DBP concentrations obtained at the different UV doses employed and their standard deviations (shown as error bars). These figures have been plotted on a log scale only to facilitate visualization of the data and comparison of trends. Data obtained without UV irradiation (i.e. at 0 mJ/cm^2 UV dose) have been plotted as 1 mJ/cm^2 UV dose to enable inclusion of this data in this style of plot.

Trihalomethanes

THMs from the Disinfection and Photolysis of Synthetic Water and UFC Testing

The differences in THM yields obtained from synthetic water irradiation with the different UV technologies alone (no peroxide present) were small, particularly in the UV dose range typical for disinfection. In the absence of hydrogen peroxide, UV irradiation followed by

UFC testing at low doses (<500 mJ/cm^2) did not have an effect on the THM concentrations observed following UFC testing (Figures 5.1a,b, Table 5.1). This was true for each of the lamp technologies tested, and for each of the laboratories. The results are in general agreement with those reported by other researchers (Malley et al. 1995, Zheng et al. 1999a,b).

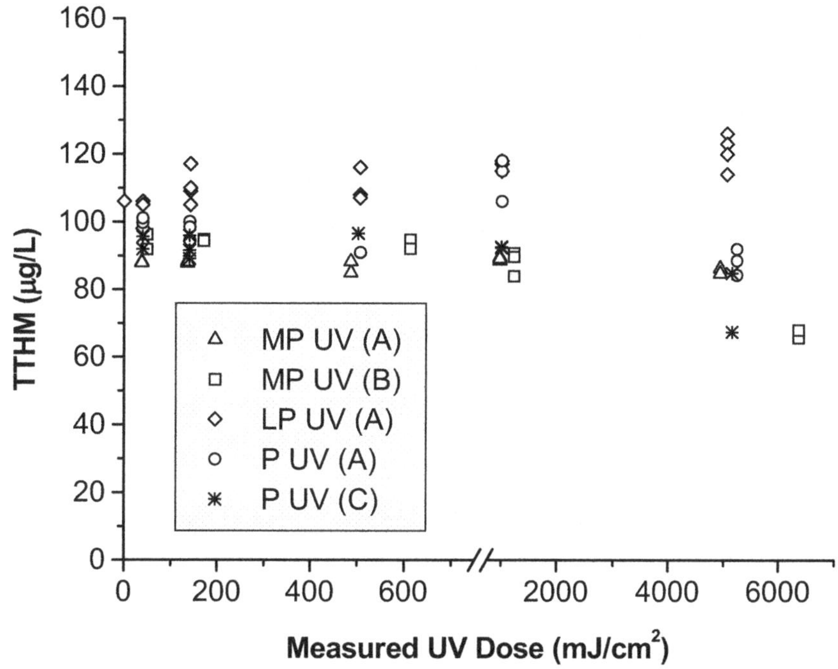

Figure 5.1a TTHM concentrations from UV irradiation of synthetic water followed by UFC testing (showing all replicates)

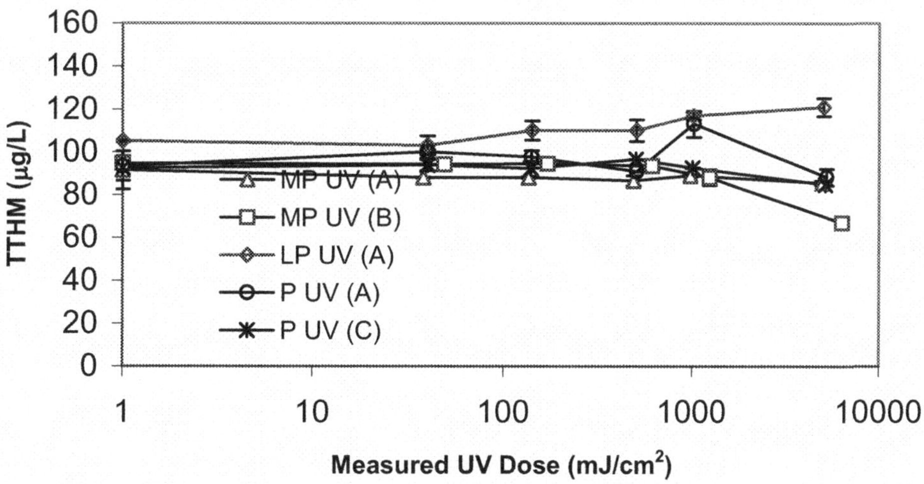

Figure 5.1b TTHM concentrations from UV irradiation of synthetic water followed by UFC testing (averages and standard deviations) (from Liu et al 2002a. Reprinted from *Water Science and Technology: Water Supply*, volume 2, issue number 5&6, pages 515-521, with permission from the copyright holders, IWA Publishing)

Table 5.1
TTHM data for the UV irradiation and UFC testing of synthetic water

MP UV (A)			MP UV (B)			LP UV (A)		
Measured UV Dose	TTHM		Measured UV Dose	TTHM		Measured UV Dose	TTHM	
(mJ/cm^2)	(μg/L)	Std.Dev.	(mJ/cm^2)	(μg/L)	Std.Dev.	(mJ/cm^2)	(μg/L)	Std.Dev.
0*	91.4	8.9	0*	94.6	NA	0*	105	0.4
38	88.0	0.2	49	94.1	3.0	40	103	4.4
136	88.0	0.3	171	94.5	0.3	142	110	4.5
486	86.6	2.3	613	93.4	1.8	506	110	5.1
978	89.0	0.5	1238	88.1	3.7	1015	117	1.8
4935	85.5	1.1	6360	66.9	1.6	5062	121	4.2

- plotted as 1 mJ/cm^2 in Figure 5.1

(Continued)

Table 5.1 (Continued)
TTHM data for the UV irradiation and UFC testing of synthetic water

P UV (A)			P UV (C)		
Measured UV Dose	TTHM		Measured UV Dose	TTHM	
(mJ/cm^2)	(μg/L)	Std.Dev.	(mJ/cm^2)	(μg/L)	Std.Dev.
0*	93.2	0.0	0*	92.3	5.8
40	100	0.9	40	93.8	2.6
141	97.6	2.9	140	92.2	3.5
507	90.9	0.1	502	96.6	NA
1020	113	6.2	1006	92.5	0.2
5244	88.2	3.9	5153	84.6	NA

* plotted as 1 mJ/cm^2 in Figure 5.1

Higher UV doses (>1000 mJ/cm^2) did not affect the resulting THM concentrations from subsequent UFC tests in the MP UV and P-UV experiments performed in Lab A and P-UV experiments performed in Lab C. However, upon increasing the UV dose to a nominal 5000 mJ/cm^2, a 15 % increase in THM concentrations was observed in LP UV testing performed in Lab A, and a 29% decrease in THM concentrations was observed in the MP UV experiments performed in Lab B.

The decrease in THMs observed in the MP UV experiments performed in Lab B was determined to be due to the presence of approximately 2 mg/L hydrogen peroxide during UV irradiation, resulting in advanced oxidation conditions. When reconsidered with this information, these results are comparable to those obtained from formal testing of the advanced oxidation process, described in the next section of this report.

The THMs were composed almost exclusively of chloroform in these samples. This was expected given that the synthetic water did not contain appreciable concentrations of bromide.

LP UV/H_2O_2, MP UV/H_2O_2 and P-UV/H_2O_2 technologies performed similarly in regards to their effect on THM concentrations resulting from the advanced oxidation of synthetic water followed by chlorination (UFC testing). While all of these AOPs did have an effect on ultimate THM concentrations, the differences in THM yields obtained from these UV/H_2O_2 technologies were small, particularly in the UV dose range typical for disinfection. However, in the P-UV experiments performed in Lab A, increases in THM concentrations were observed to be 50% and 46% at the nominal doses of 500 and 1000 mJ/cm^2, respectively. The reason for this abnormality was not identified in this research.

For UV irradiation in the presence of nominally 100 mg/L H_2O_2 followed by UFC testing, there was a direct relationship between UV dose and observed THM concentrations for UV doses of up to approximately 500 mJ/cm^2 (Figures 5.2a,b,c and Table 5.2). This was true for both MP (maximum 9% to 11 % increase) and LP (maximum 19 % increase) UV/ H_2O_2 technologies. In contrast, there was either an abrupt increase by approximate 50% (Lab A) or fluctuating variation (Lab C) in TTHM concentrations by P-UV/H_2O_2 irradiation with the same doses range as above. Higher UV doses (>1000 mJ/cm^2) resulted in decreases in THM concentrations from subsequent UFC tests with all UV technologies and in all laboratories. Reductions of 22 % and 50 % were observed for the advanced oxidation performed with MP UV doses of 5202 and 6415 mJ/cm^2, respectively, and a 50 % decrease was observed for LP UV dosed at 5150 mJ/cm^2. 36% and 44% reductions in THM concentrations were observed for P-UV/H_2O_2 irradiation with the doses of 5576 and 5710 mJ/cm^2.

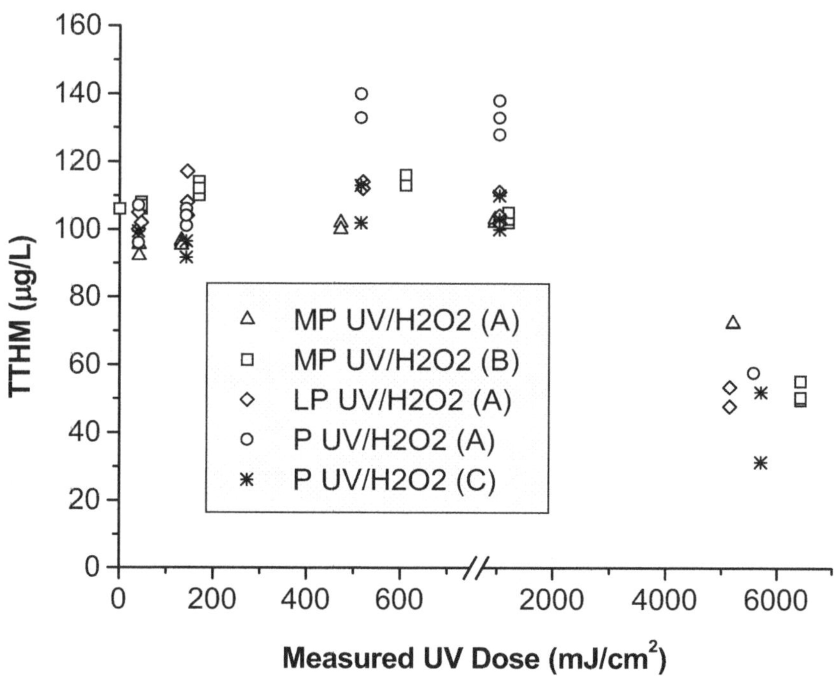

Figure 5.2a TTHM concentrations from UV irradiation of synthetic water in the presence of nominally 100 mg/L H_2O_2, followed by UFC testing (showing all replicates)

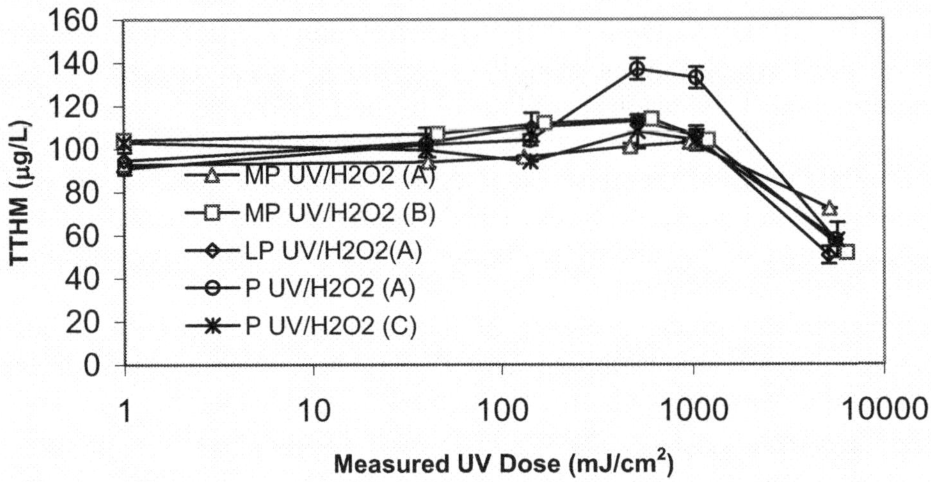

Figure 5.2b TTHM concentrations from UV irradiation of synthetic water in the presence of nominally 100 mg/L H_2O_2, followed by UFC testing (averages and standard deviations) (Source: Liu et al. 2002b. Reprinted from 2002 *AWWA Water Quality Technology Conference Proceedings* by permission. Copyright 2002, American Water Works Association).

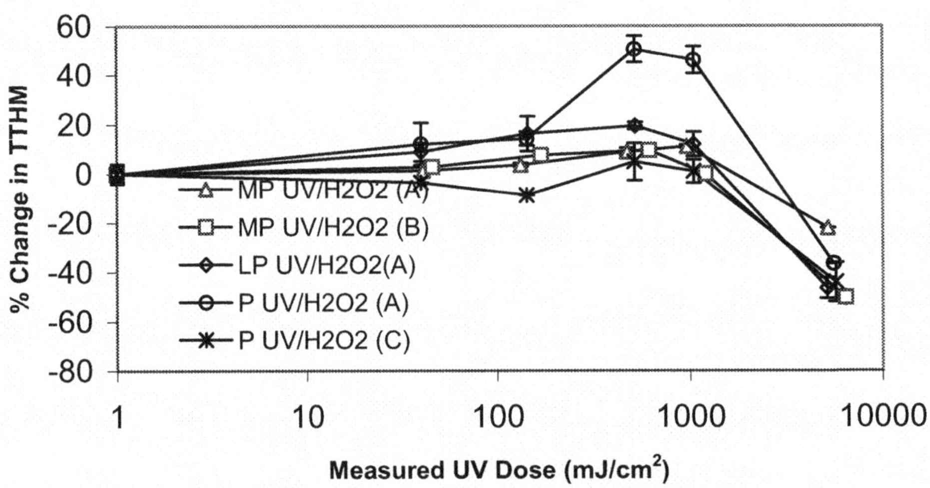

Figure 5.2c Changes in TTHM concentrations from UV irradiation of synthetic water in the presence of nominally 100 mg/L H_2O_2, followed by UFC testing

Table 5.2
TTHM data for the UV/H_2O_2 irradiation and UFC testing of synthetic water

MP UV/H_2O_2 (A)			MP UV/H_2O_2 (B)			LP UV/H_2O_2 (A)		
Measured UV Dose	TTHM		Measured UV Dose	TTHM		Measured UV Dose	TTHM	
(mJ/cm^2)	(µg/L)	Std.Dev.	(mJ/cm^2)	(µg/L)	Std.Dev.	(mJ/cm^2)	(µg/L)	Std.Dev.
0*	92.7	1.6	0*	104	1.7	0*	94.6	3.8
41	93.9	2.3	46	107	1.0	40	103	3.5
132	96.1	0.7	169	112	1.7	144	110	6.7
472	101	1.4	611	114	1.5	519	113	1.4
964	103	0.6	1203	104	1.3	1044	106	4.7
5202	72.3	0.1	6415	51.8	3.1	5150	50.7	4.0

* plotted as 1 mJ/cm^2 in Figures 5.2 and 5.3; Nominally 100 mg/L H_2O_2

(Continued)

Table 5.2 (Continued)
TTHM data for the UV/H_2O_2 irradiation and UFC testing of synthetic water

P-UV/H_2O_2 (A)			P-UV/H_2O_2 (C)		
Measured UV Dose	TTHM		Measured UV Dose	TTHM	
(mJ/cm^2)	(µg/L)	Std.Dev.	(mJ/cm^2)	(µg/L)	Std.Dev.
0*	91.0	2.9	0*	103	0.1
40	102	7.8	40	99.4	NA
142	104	2.5	142	94.1	NA
515	137	4.9	515	108	7.8
1045	133	5.0	1045	104	4.9
5576	57.8	NA	5710	57.7	8.1

* plotted as 1 mJ/cm^2 in Figures 5.2 and 5.3
Nominally 100 mg/L H_2O_2

While direct comparison of UV doses employed in the experiments of other researchers is difficult, the general trends of these results are in agreement with those reported by other researchers (Zheng et al., 1999a,b; Kleiser and Frimmel, 2000). Kleiser and Frimmel (2000) suggested that lower reaction rates observed when using 'higher' peroxide concentrations (16 mg/L) could be due to a partial scavenging effect of the peroxide. Zheng and co-workers and Speitel and co-workers (Speitel and Symons 1999) found that reaction rates tended to increase with increasing peroxide concentration until a plateau was reached (e.g. at approximately 25 mg/L peroxide in the Zheng work).

Additional experiments were performed to better define the behavior of these systems at high UV doses, and to investigate the potential role of hydrogen peroxide in the UV AOP process. Doses of approximately 0, 1000, 2000, 3000, 4000, and 5000 mJ/cm^2 were used in the experiments, and initial H_2O_2 concentrations were 98.8 mg/L, when applicable. The results, described in detail in Appendix C, further demonstrated that precursors of TTHMs were partly destroyed by UV irradiation at high UV doses, and the presence of H_2O_2 enhanced the destruction of TTHM precursors.

Again, the THMs were composed almost exclusively of chloroform in all of these samples. This was not unexpected given that the synthetic water did not contain appreciable concentrations of bromide.

Reproducibility of UV Irradiation-UFC Testing for TTHM Generation

The UFC test was reasonably reproducible, with standard deviations of replicate samples of generally less than 5 μg/L being obtained for TTHM concentrations of 51 to 137 μg/L.

In regards to differences observed for samples from different laboratories, UFC testing of synthetic water prepared in each of Labs A, B and C (i.e. at a UV dose of 0 mJ/cm^2) showed only minor differences in the THM concentrations observed (91.4 to 105 μg/L obtained before UV irradiation). This indicated that both the preparation of the synthetic water and UFC testing procedure were consistently performed with respect to the ultimate formation of THMs.

Both the photolysis and advanced oxidation procedures were performed comparably well in Labs A, B and C, as indicated by the similarity in the UFC THM concentrations obtained by experiments performed in Labs A and B in parallel involving MP UV and MP UV/H_2O_2. This is especially true for the MP UV/H_2O_2 experiments. Both laboratories reported a 9 to 11 % increase in THM formation at a nominal UV dose of 500 mJ/cm^2, and the decreases in THM concentrations observed at higher doses were consistent with the differences in the applied UV doses. Similarly, the results of both laboratories were comparable to each other for MP UV applied at doses of < 500 mJ/cm^2. As was stated previously, differences in the results at higher MP UV doses are likely attributable to the presence of small amounts of hydrogen peroxide in the synthetic water matrix of Lab B in these experiments.

Haloacetic Acids

HAAs from the Disinfection and Photolysis of Synthetic Water and UFC Testing

In the absence of hydrogen peroxide, LP UV irradiation followed by UFC testing did not have an effect on the HAA concentrations observed regardless of the UV dose employed (Figures 5.3a,b).

The trend in HAA formation following MP UV irradiation and UFC testing was similar to that observed for THM formation under advanced oxidation conditions (previous section). HAA concentrations increased with increasing UV dose to a maximum value at approximately 140 mJ/cm^2, then decreased with further increases in UV dose. Maximum observed increases in HAA concentration were 9 and 34 % for Labs A and B, respectively, while the greatest decreases in HAA concentration (at the highest UV doses) were observed to be 15 and 43 %, respectively.

The P-UV technologies did not have appreciable effects on the HAA concentration generated in either Lab A or lab C, although slight decreases of 8% and 15% were observed at high UV doses of 5244 mJ/cm^2 and 5153 mJ/cm^2, respectively.

The HAAs were composed almost exclusively of dichloroacetic acid and trichloracetic acid in these samples. This was expected given that the synthetic water did not contain appreciable concentrations of bromide.

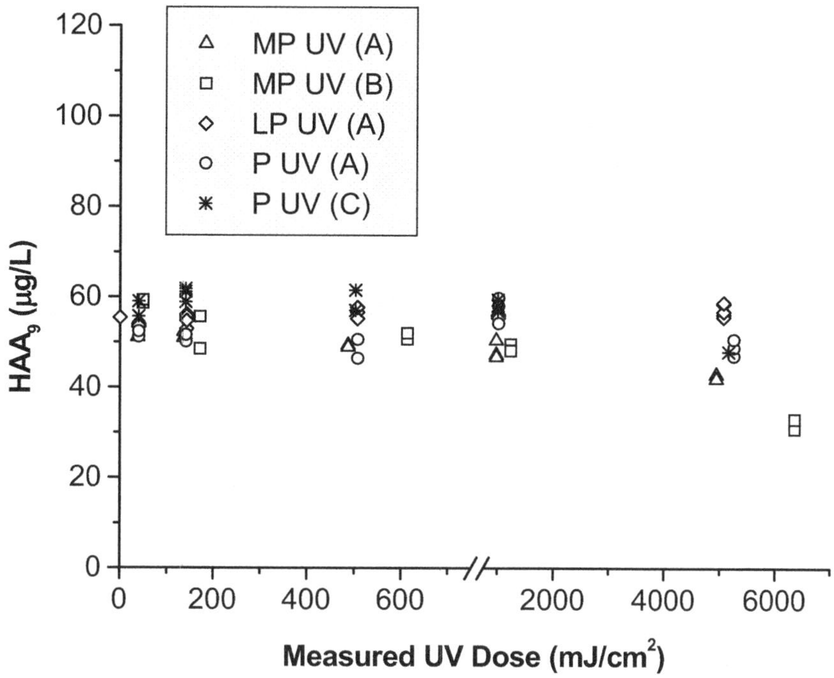

Figure 5.3a HAA$_9$ concentrations from UV irradiation of synthetic water followed by UFC testing (showing all replicates)

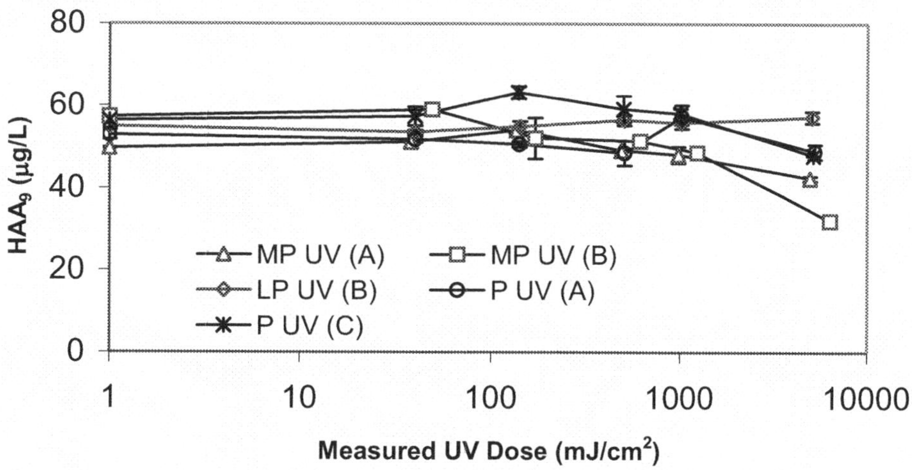

Figure 5.3b HAA$_9$ concentrations from UV irradiation of synthetic water followed by UFC testing (averages and standard deviations) (from Liu et al 2002a. Reprinted from *Water Science and Technology: Water Supply*, volume 2, issue number 5&6, pages 515-521, with permission from the copyright holders, IWA Publishing)

The differences in HAA yields obtained from the different UV/H_2O_2 technologies were small throughout the UV dose range tested, and the effect of UV dose was similar to that observed for THM formation (described in a previous section).

For UV irradiation in the presence of nominally 100 mg/L H_2O_2 followed by UFC testing, there was a small but direct relationship between UV dose and observed HAA concentrations for UV doses of up to approximately 500 mJ/cm^2 (Figures 5.4a,b, and 5.5). This was true for each of the MP (maximum 5 to 9 % increase), LP UV (maximum 9 % increase) and P-UV technologies (maximum 25 to 7%), and for each of Labs A, B and C. Higher UV doses (>1000 mJ/cm^2) resulted in decreases in HAA concentrations from subsequent UFC tests with all UV technologies and in all laboratories. Reductions of 44 % and 59 % were observed for the advanced oxidation performed with MP UV doses of 5202 and 6415 mJ/cm^2, respectively, and a 60 % decrease was observed for LP UV dosed at 5150 mJ/cm^2. Similarly, P-UV/H_2O_2 irradiation resulted in 53% and 65% reductions in HAA concentrations with doses of 5576 mJ/cm^2 and 5710 mJ/cm^2, respectively.

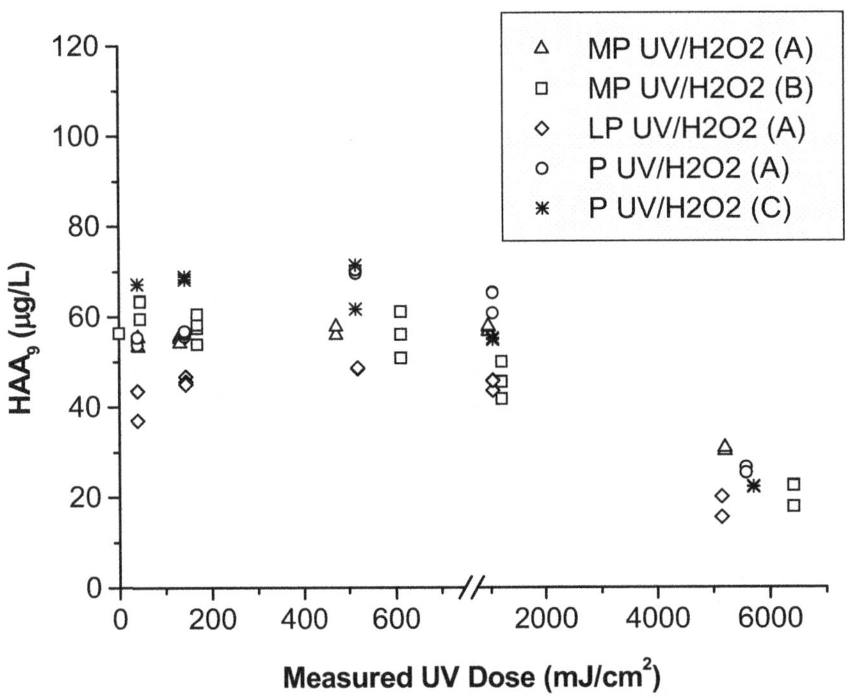

Figure 5.4a HAA$_9$ concentrations from UV irradiation of synthetic water in the presence of nominally 100 mg/L H_2O_2, followed by UFC testing (showing all replicates)

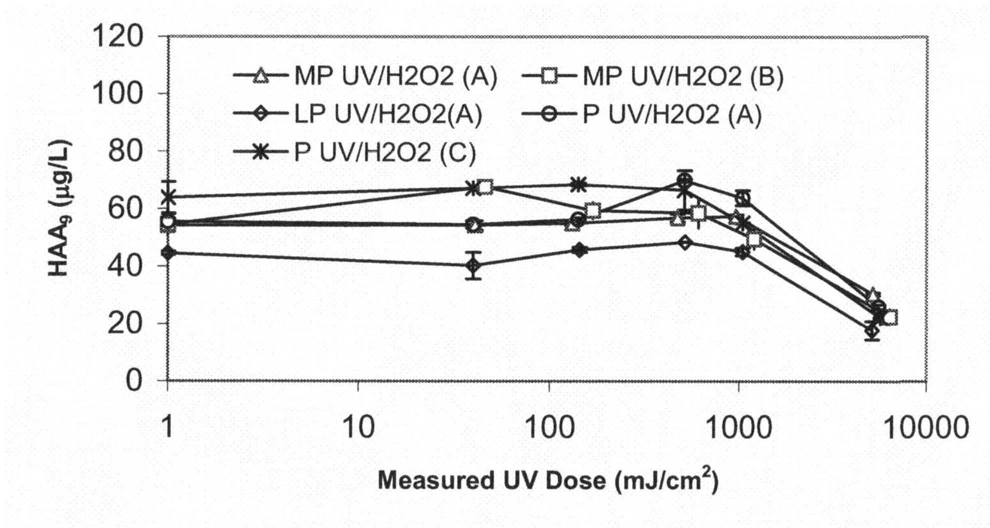

Figure 5.4b HAA$_9$ concentrations from UV irradiation of synthetic water in the presence of nominally 100 mg/L H_2O_2, followed by UFC testing (averages and standard deviations) (Source: Liu et al. 2002b. Reprinted from 2002 *AWWA Water Quality Technology Conference Proceedings* by permission. Copyright 2002, American Water Works Association)

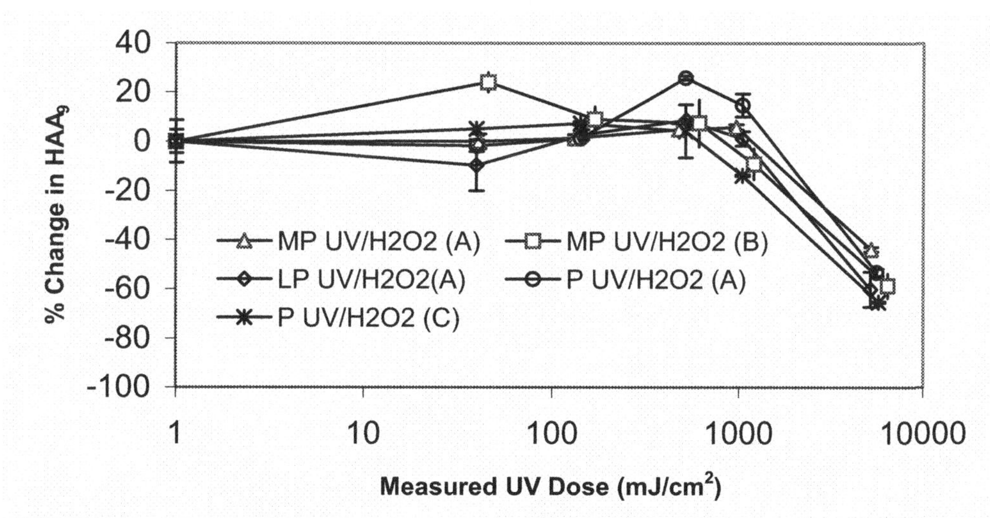

Figure 5.5 Changes in HAA$_9$ concentrations from UV irradiation of synthetic water in the presence of nominally 100 mg/L H_2O_2, followed by UFC testing

Again, the HAAs were composed almost exclusively of dichloroacetic acid and trichloracetic acid in these samples. This was expected given that the synthetic water did not contain appreciable concentrations of bromide.

Reproducibility of UV Irradiation-UFC Testing for HAA Generation

The UFC test was reasonably reproducible, with standard deviations of replicate samples of generally less than 3 µg/L being obtained for HAA data of concentrations of 32 to 70 µg/L.

In regards to differences observed for samples from different laboratories, UFC testing of synthetic water prepared in each of Labs A and B (i.e. at a UV dose of 0 mJ/cm^2) showed only minor differences in the HAA concentrations observed (49 to 56 µg/L obtained before UV irradiation). Only one synthetic water sample generated by Lab C had a slightly higher HAA concentration of 63.9 µg/L, although it was still reasonably consistent with other samples prepared by the other labs. In short, most of the data indicated that both the preparation of the synthetic water and UFC testing procedure were generally consistently performed with respect to the ultimate formation of HAAs.

Both the photolysis and advanced oxidation procedures were performed comparably well in Labs A, B and C, as indicated by the similarity in HAA results obtained by all labs in parallel experiments involving MP UV and MP UV/H_2O_2. This is especially true for the MP UV/H_2O_2 experiments. Both laboratories reported a 5 to 9 % increase in HAA formation at a nominal UV dose of 500 mJ/cm^2, and the decreases in HAA concentrations observed at higher doses were consistent with the differences in the applied UV doses.

Aldehydes

Aldehydes from the Disinfection and Photolysis of Synthetic Water

Aldehydes were produced from each of LP UV, MP UV and P-UV irradiation, although significant increases in concentrations above background levels were not observed below applied doses of approximately 500 mJ/cm^2 for either UV technology (Figures 5.6a,b). Above 500 mJ/cm^2, a direct relationship between UV dose and aldehyde concentration was observed under irradiation of MP UV and P-UV, although increases in aldehyde concentrations with increases in LP UV dose were only approximately half those observed for MP UV irradiation of synthetic water. For P-UV irradiation, noticeable increases in aldehydes were only observed with the dose above 1000 mJ/cm^2. Maximum increases in aldehyde concentrations observed at the highest UV doses applied were approximately 13 µg/L for LP UV irradiation, 29 to 45 µg/L for MP UV irradiation, and 18 to 31 µg/L for P-UV irradiation. Although a relatively high level of aldehydes (83 µg/L) was observed in the synthetic water of Lab B, this did not appear to affect the formation of additional aldehydes upon UV irradiation.

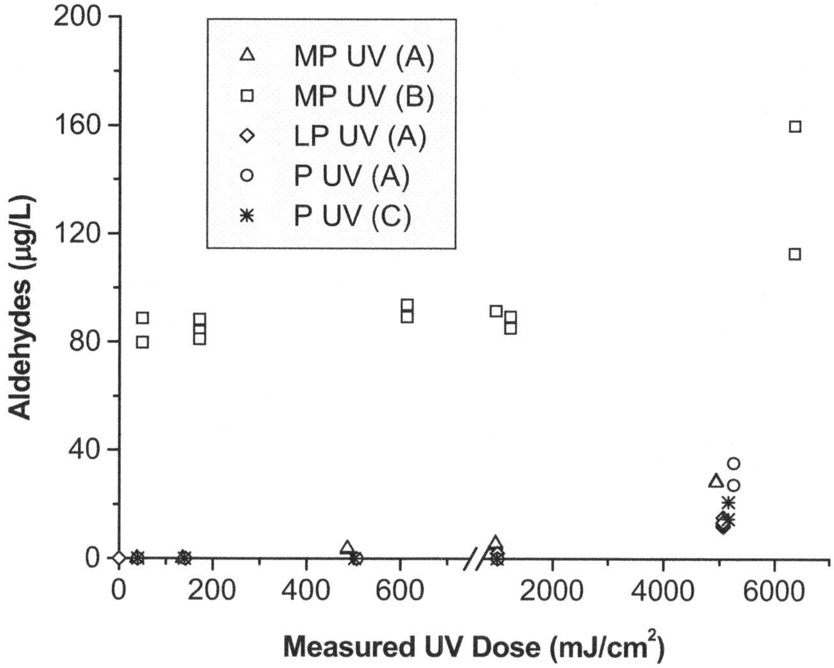

Figure 5.6a Aldehyde concentrations from UV irradiation of synthetic water (showing all replicates)

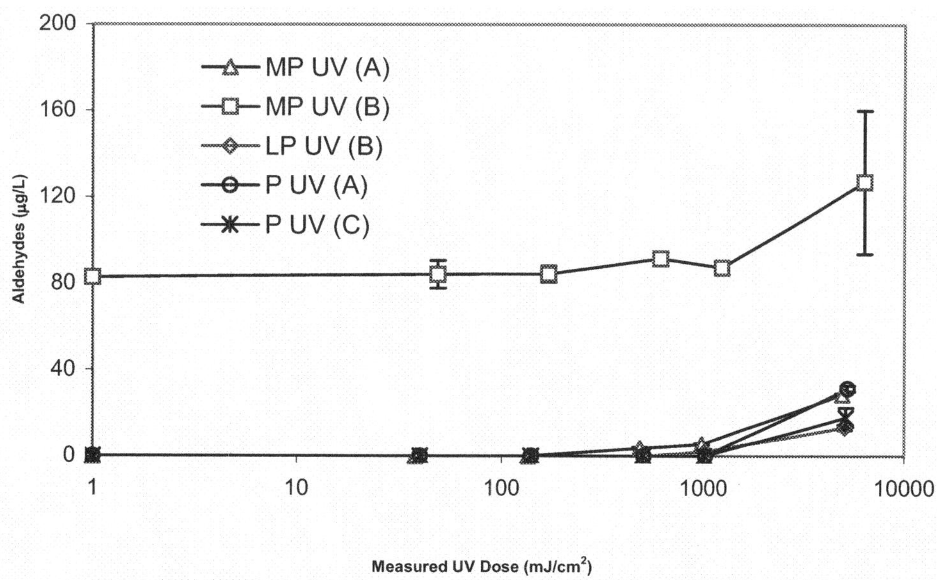

Figure 5.6b Aldehyde concentrations from UV irradiation of synthetic water (averages and standard deviations) (from Liu et al 2002a. Reprinted from *Water Science and Technology: Water Supply*, volume 2, issue number 5&6, pages 515-521, with permission from the copyright holders, IWA Publishing)

Aldehydes were produced from all LP UV/H_2O_2, MP UV/H_2O_2 and P-UV/H_2O_2 irradiation, although significant increases in concentrations above background levels were not observed below applied doses of approximately 140 mJ/cm^2 for either MP/H_2O_2 or LP/H_2O_2 UV technology (Figures 5.7a,b). While for P-UV/H_2O_2 technology, the threshold for noticeable aldehyde yields was nominally 1000 mJ/cm^2. Above 140 mJ/cm^2 for MP/H_2O_2 or LP UV/H_2O_2, or above 1000 mJ/cm^2 for P-UV/H_2O_2, a direct relationship between UV dose and aldehyde concentration was generally observed. However, increases in aldehyde concentrations with increases in LP UV/H_2O_2 or P-UV/H_2O_2 dose were only approximately half those observed for MP UV/H_2O_2 irradiation of synthetic water. For the three experiments performed in Lab A, maximum increases in aldehyde concentrations observed at the highest UV doses applied were approximately 21 µg/L for LP UV/H_2O_2 irradiation, 41 µg/L for MP UV/H_2O_2 irradiation and 17.8 µg/L for P-UV/H_2O_2 irradiation. While 22 µg/L was observed at the highest UV dose applied for P-UV/H_2O_2 irradiation conducted in Lab C.

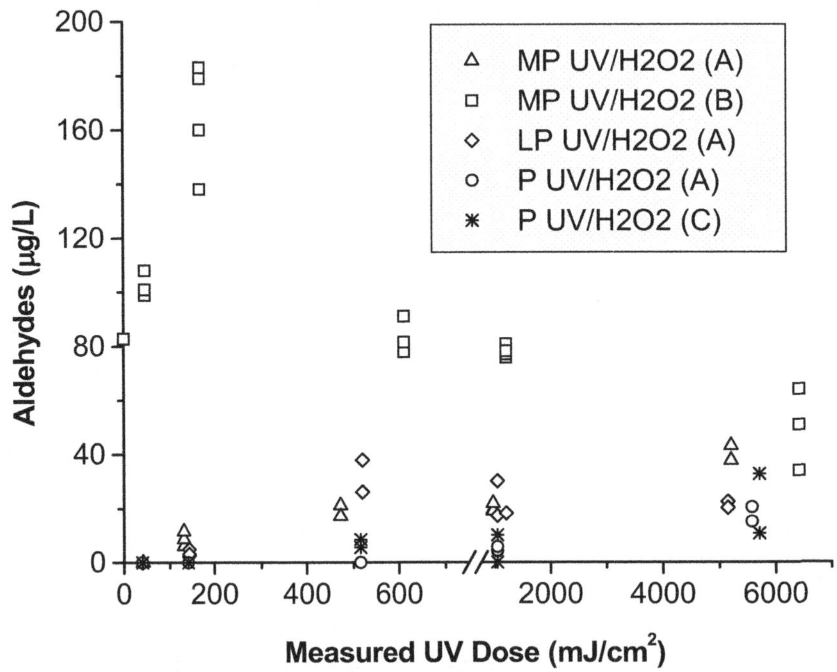

Figure 5.7a Aldehyde concentrations from UV irradiation of synthetic water in the presence of nominally 100 mg/L H_2O_2, (showing all replicates)

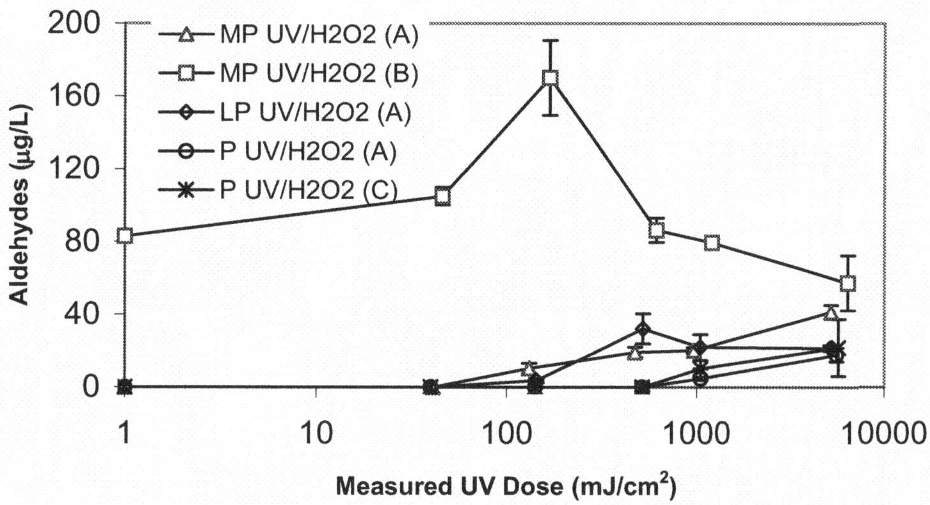

Figure 5.7b Aldehyde concentrations from UV irradiation of synthetic water in the presence of nominally 100 mg/L H_2O_2, (averages and standard deviations) (Source: Liu et al. 2002b. Reprinted from 2002 *AWWA Water Quality Technology Conference Proceedings* by permission. Copyright 2002, American Water Works Association).

The exception was the experiment performed using MP UV/H_2O_2 in Lab B, in which the trends were similar to those observed for THM and HAA formation described in previous sections (i.e. initial increases in aldehyde followed by later decreases). Approximately 2 mg/L hydrogen peroxide and a relatively high level of aldehydes (83 µg/L) were observed in the synthetic water prior to UV irradiation. There may have been other factors that caused the different results between the two labs for the MP UV/H_2O_2 experiments.

Reproducibility of UV Irradiation for Aldehyde Generation

Experiments on the MP UV irradiation of synthetic water in Labs A, B and C showed that the reproducibility of the irradiation and analytical procedures was good, resulting in relative standard deviations in replicate tests of less than 5%. Also, similar changes in aldehyde yields were observed in each laboratory. This was despite the relatively large background concentration of aldehydes in the synthetic water matrix of Lab B.

In the MP UV/H_2O_2 experiments, the reproducibility of the irradiation and analytical procedures was good within each laboratory and between Lab A and C, however the high background aldehyde concentrations in Lab B prevented direct comparison of absolute aldehyde concentrations. Additional water quality differences associated with the high aldehyde background concentrations may have also caused the differences in general trends observed for the effect of UV dose on aldehyde concentrations between the Lab A and B.

Carboxylic Acids

Carboxylic Acids from the Disinfection and Photolysis of Synthetic Water

Carboxylic acids appeared to be produced from all LP UV, MP UV and P-UV irradiation, although significant increases in concentrations above background levels were not observed below applied doses of approximately 1000 mJ/cm^2 for either MP UV or LP UV technology (Figures 5.8a,b). However, a significant increase in carboxylic acid concentrations was observed with the P-UV doses above 40 mJ/cm^2 performed in Lab C. Also, in the case of the LP UV and P-UV experiments performed in Lab A, the scatter in the data was such that changes in carboxylic acid concentration may not have been statistically significant for that UV technology. Maximum increases in carboxylic acid concentrations observed at the highest UV doses applied were approximately 100 µg/L for MP UV experiments performed at Lab A, 240 µg/L for MP UV experiments performed at Lab B and 256 µg/L for P-UV experiments performed in Lab C. Although a relatively high level of carboxylic acids (213 µg/L) was observed in the synthetic water of Lab B, this did not appear to affect the formation of additional carboxylic acids upon UV irradiation.

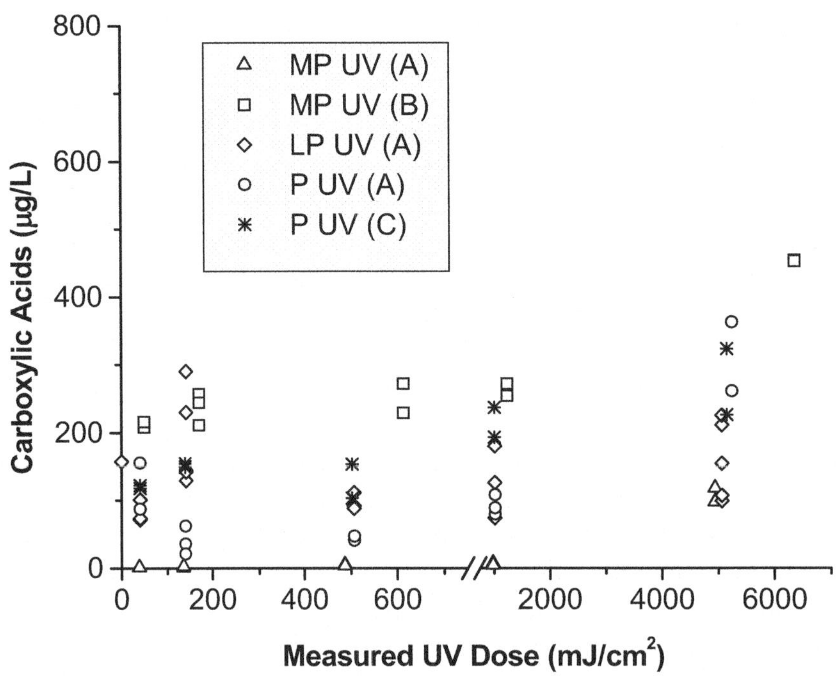

Figure 5.8a Carboxylic acid concentrations from UV irradiation of synthetic water (showing all replicates)

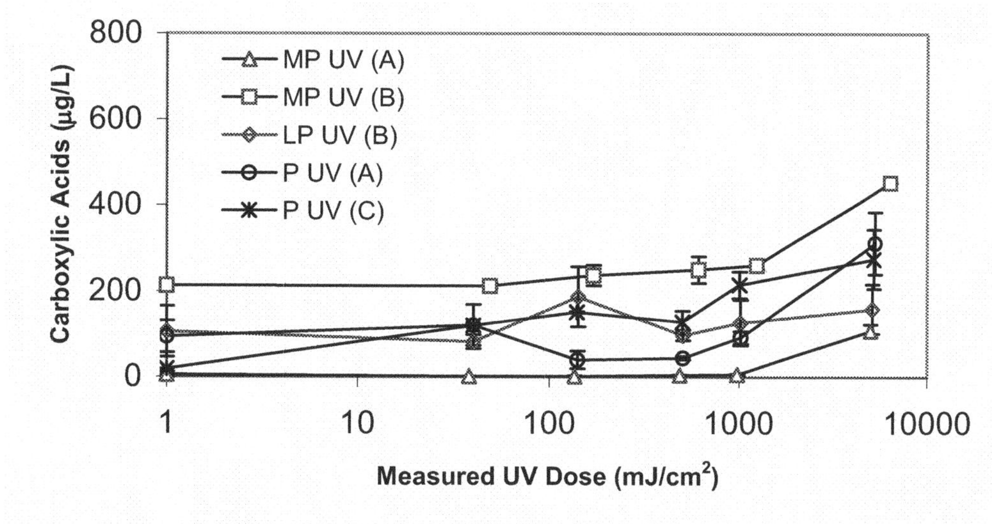

Figure 5.8b Carboxylic acid concentrations from UV irradiation of synthetic water (averages and standard deviations) (from Liu et al 2002a. Reprinted from *Water Science and Technology: Water Supply*, volume 2, issue number 5&6, pages 515-521, with permission from the copyright holders, IWA Publishing)

Carboxylic Acids from the Advanced Oxidation of Synthetic Water

Carboxylic acids were produced from all LP UV/H_2O_2, MP UV/H_2O_2 and P-UV irradiation, although significant increases in concentrations above background levels were not observed below applied doses of approximately 140 mJ/cm^2 for either UV technology (Figures 5.9a,b). Under irradiation of LP UV/H_2O_2 performed by Lab A and MP-UV/H_2O_2 performed by Lab B with the doses above 140 mJ/cm^2, a direct relationship between UV dose and aldehyde concentration was generally observed up to approximately 1000 mJ/cm^2 when the concentrations appeared to plateau. For the P-UV technology, a continued increase was observed even at the doses above 1000 mJ/cm^2, hence no plateau was observed to exist. An exception was the experiment performed using MP UV/H_2O_2 in Lab A, in which no carboxylic acids had been observed.

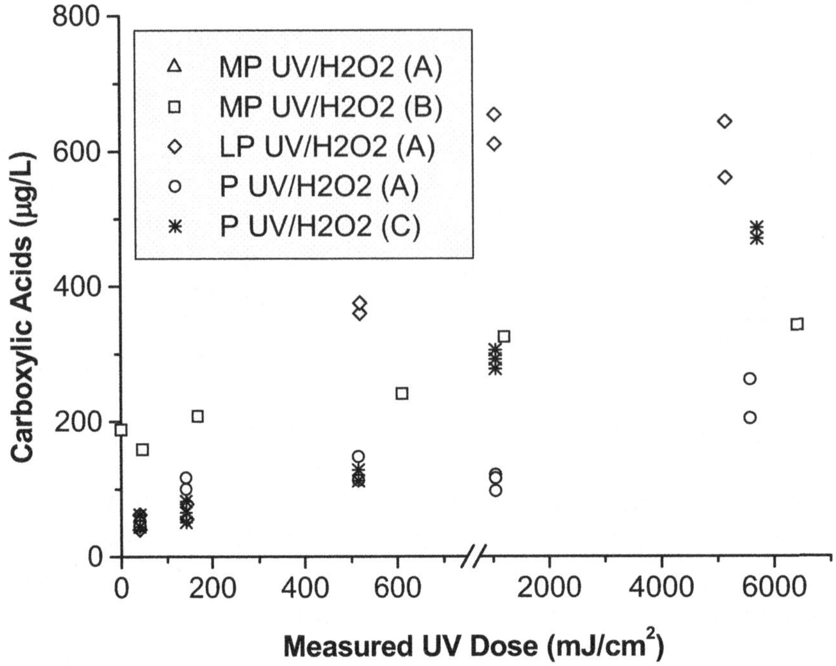

Figure 5.9a Carboxylic acid concentrations from UV irradiation of synthetic water in the presence of nominally 100 mg/L H_2O_2, (showing all replicates)

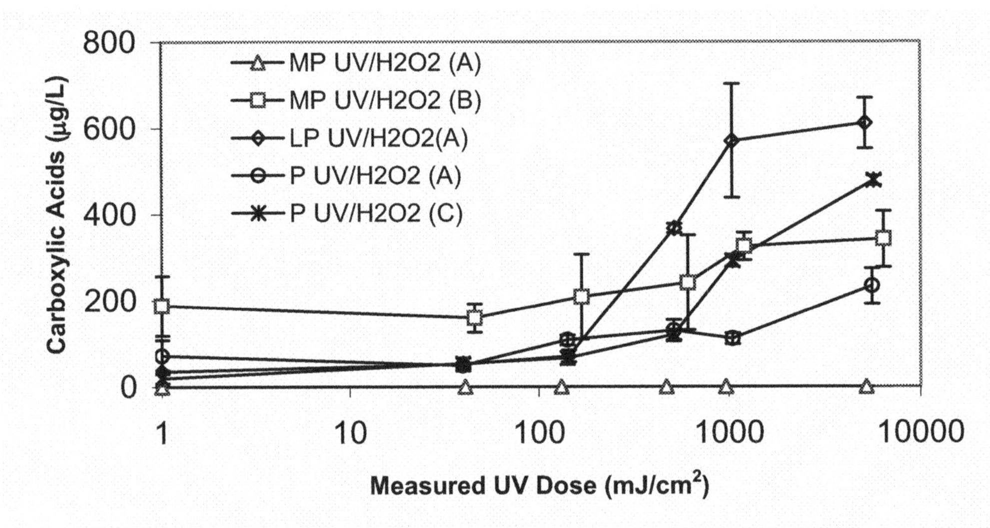

Figure 5.9b Carboxylic acid concentrations from UV irradiation of synthetic water in the presence of nominally 100 mg/L H_2O_2, (averages and standard deviations) (Source: Liu et al. 2002b. Reprinted from 2002 *AWWA Water Quality Technology Conference Proceedings* by permission. Copyright 2002, American Water Works Association).

Contrary to observations for other DBPs, the greatest increases in carboxylic acid concentration were observed for the LP UV/H_2O_2 system. Maximum increases in carboxylic acid concentrations observed at the highest UV doses applied were approximately 580 μg/L for LP UV/H_2O_2 irradiation, and 160 μg/L for MP UV/H_2O_2 irradiation.

Reproducibility of UV Irradiation for Carboxylic Acid Generation

The carboxylic acid analytical procedure carries with it the lowest precision of the four classes of DBPs included in this project (RSD of 2.9 to 47 %). The higher RSDs have typically been associated with earlier experiments in this project, with RSDs values of approximately 10 to 20 % being obtained for more recent experiments. Despite this and the relatively high background levels of carboxylic acids present in the synthetic water matrix, the effects of UV irradiation on carboxylic acid formation may still be observed. For example, the relatively high level of carboxylic acids (213 μg/L) observed in the synthetic water of Lab B did not appear to affect the observation of the formation of additional carboxylic acids upon MP UV irradiation. Nor did the presence of 35 μg/L and 188 μg/L in the synthetic water of Labs A and B, respectively, during the AOP testing.

EFFECTS OF DIFFERENT WATER TYPES ON DBP FORMATION FROM MP-UV AND MP-UV/H_2O_2

Selected MP-UV irradiations were repeated at bench-scale and pilot-scale using filtered Grand River water, obtained from the Mannheim Water Treatment Plant, Kitchener ON Canada, instead of synthetic water to determine if the same trends would be observed in natural water as were observed with the synthetic water. In addition, the experiments were also repeated in synthetic water which had been supplemented with bromide to simulate a bromide-containing source water. The bench-scale experiments were performed using the collimated beam apparatus in the University of Waterloo laboratory. The pilot-scale irradiations were performed with the Rayox unit located at the Mannheim Water Treatment Plant, Kitchener ON Canada. MP-UV was the lamp technology of choice for these tests because MP-UV generally showed greater effects on DBPs than did LP-UV.

Bench-Scale and Pilot-Scale Testing with Filtered River Water

The general trends in the results obtained for filtered Grand River water were similar to those obtained with synthetic water, regardless of the scale at which they were obtained (bench-scale or pilot-scale). While it was found that the synthetic water did not model the natural water perfectly in regards to absolute concentrations of DBPs, the trends relating to changes in DBP concentrations in response to changes in UV doses were modeled reasonably well. Also, the distribution of individual DBP species obtained for filtered Grand River water was similar to that obtained for synthetic water, likely because the Grand River generally contains negligible concentrations of bromide. A summary of the THM results obtained is illustrated here in Figures 5.10 and 5.11.

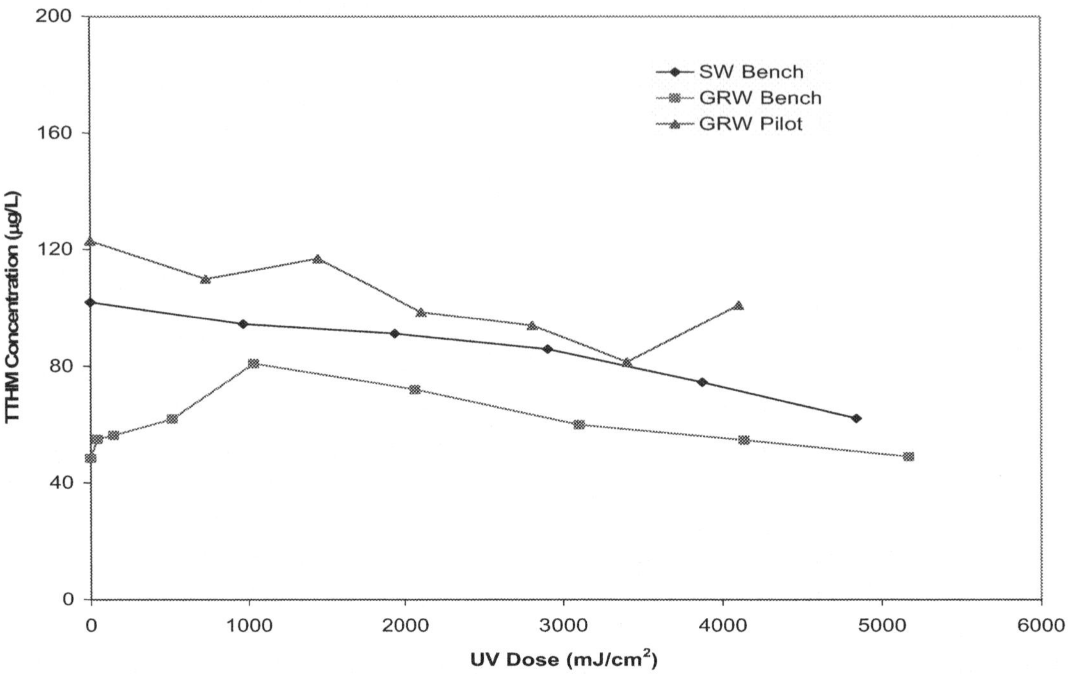

Figure 5.10 Comparison of TTHM results obtained from UV/UFC testing synthetic water (SW) and filtered Grand River Water (GRW) at bench-scale and pilot-scale

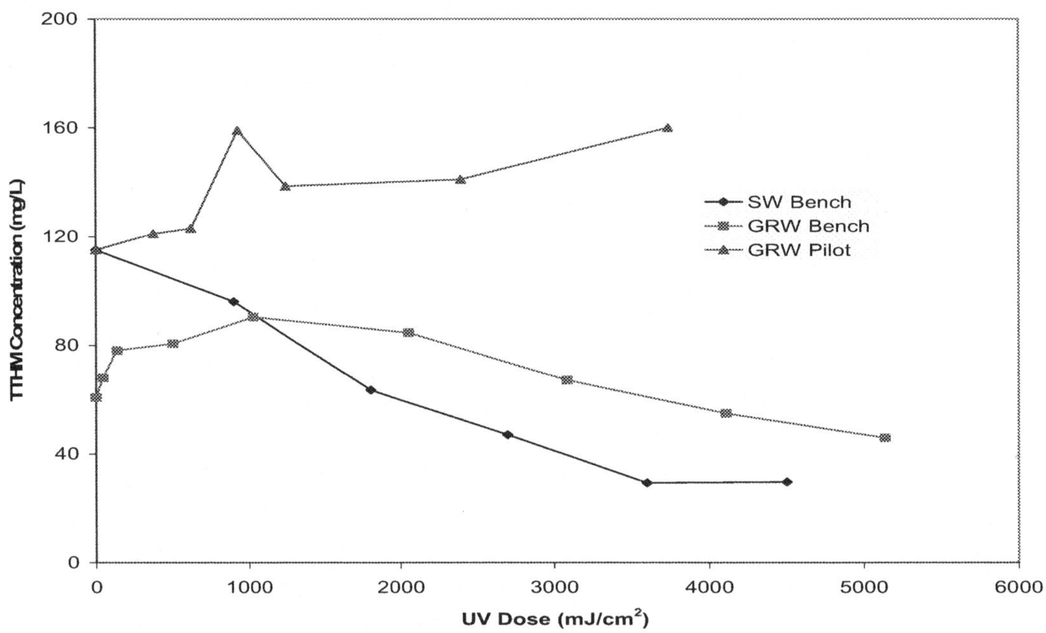

Figure 5.11 Comparison of TTHM results obtained from UV/H_2O_2/UFC testing synthetic water (SW) and filtered Grand River Water (GRW) at bench-scale and pilot-scale

Bench Testing with Filtered River Water Spiked with Bromide

Although, adding 128 μg/L NaBr to the synthetic water had the expected result of producing more brominated DBPs, and the brominated DBPs were produced in relative amounts that were typical to those observed by other researchers, trends in DBP formation relative to changes in UV dose were generally similar to those observed in the absence of bromide (i.e. in synthetic water alone). The results for each individual DBP species that was detected are shown graphically in Figures 5.12 through 5.19.

Trends in THM formation relative to changes UV dose were similar to those obtained for synthetic water alone and without the presence of hydrogen peroxide. Little effect was observed until the UV dose was increased to above approximately 1000 mJ/cm^2, beyond which the THMs from post-UV UFC testing increased with increasing UV dose (Figure 5.12) and then either leveled off or began to decrease. When the samples were irradiated under advanced oxidation conditions (~100 mg/L H_2O_2; Figure 5.13), once the UV dose was raised to above 1000 mJ/cm^2 the THM concentrations were observed to steadily decrease.

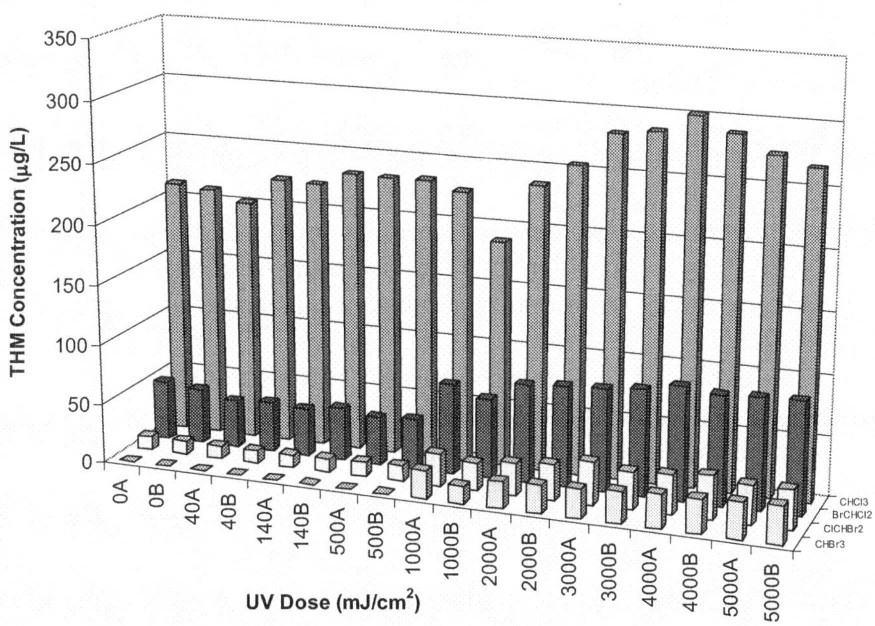

Figure 5.12 THM concentrations from UV irradiation of synthetic water containing 128 μg/L NaBr followed by UFC testing (showing all species detected, A/B replicate runs)

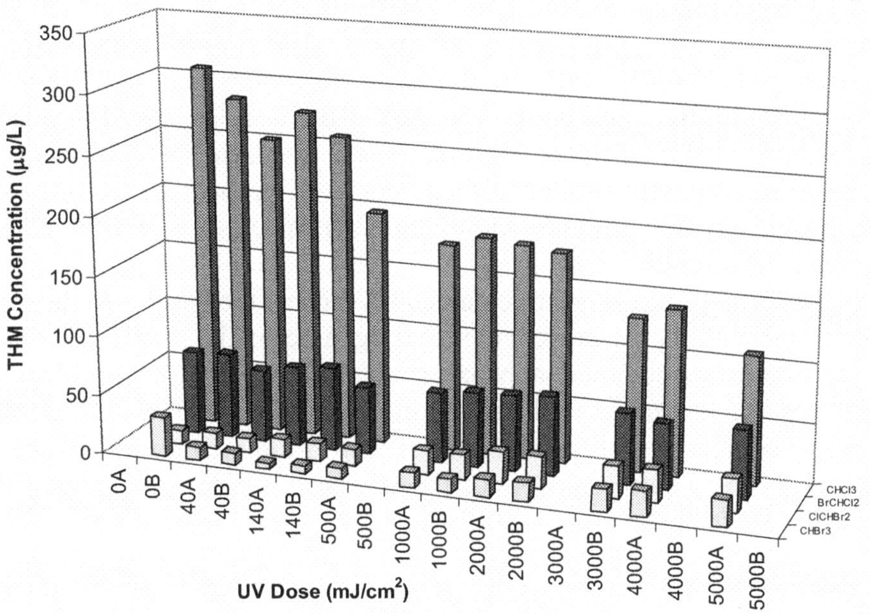

Figure 5.13 THM concentrations from UV irradiation of synthetic water in the presence of nominally 128 µg/L NaBr and 100 mg/L H_2O_2, followed by UFC testing (showing all species detected, A/B replicate runs, space=lost sample)

As was observed in the tests performed in the absence of bromide, DCAA and TCAA were still the most prevalent HAA species in these tests. In the samples to which nominally 100 µg/L bromide were added, however, the mixed chlorobromo species were also present in relatively high concentration.

Most HAA species exhibited similar responses to changes in UV dose, and these changes were consistent with those observed in synthetic water without added bromide. Under photolytic conditions, HAA concentrations increased with increasing UV doses to approximately 500 to 1000 mJ/cm^2 (Figure 5.14). Further increases in UV dose resulted in HAA concentration decreases. Under advanced oxidation conditions (nominally 100 mg/L H_2O_2), there was a steady decrease in HAA concentrations as UV doses were increased (Figure 5.15). Comparison of the different DBP species to each other under either scenario revealed that while most species exhibited similar behavior in regards to the overall trends observed, the magnitudes and thresholds of the responses were unique to each species.

Aldehyde and carboxylic acid concentrations displayed similar responses to changes in UV dose regardless of the presence or absence of bromide (Figures 5.16-5.19). Under photolytic conditions, changes in the total aldehyde concentrations were driven by changes in the acetaldehyde concentration, the formaldehyde concentration remaining relatively constant even for the highest UV doses (Figure 5.16). It is possible that there was a small amount of formaldehyde contamination in these samples, although it was not detected in the standards or blanks.

In regards to carboxylic acids, under photolysis conditions significant concentrations of most species only began to be observed once the UV dose was raised to 500 mJ/cm^2 or higher (Figure 5.18), which is consistent with that observed in the absence of bromide. The same was also true for most species under advance oxidation conditions, although formate and acetate showed continual increases with increasing UV dose even at UV doses less than 500 mJ/cm^2.

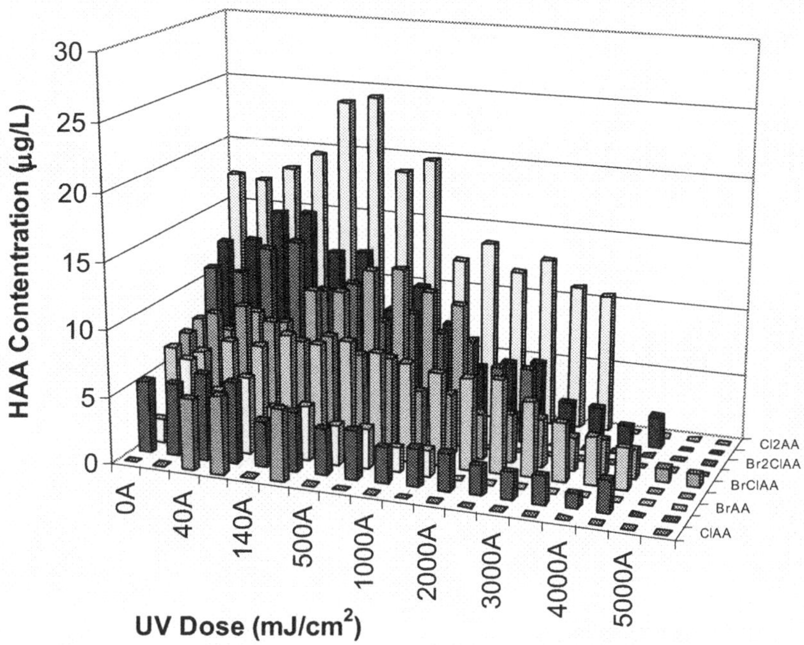

Figure 5.14 HAA concentrations from UV irradiation of synthetic water containing 128 µg/L NaBr followed by UFC testing (showing all species detected, A/B replicate runs)

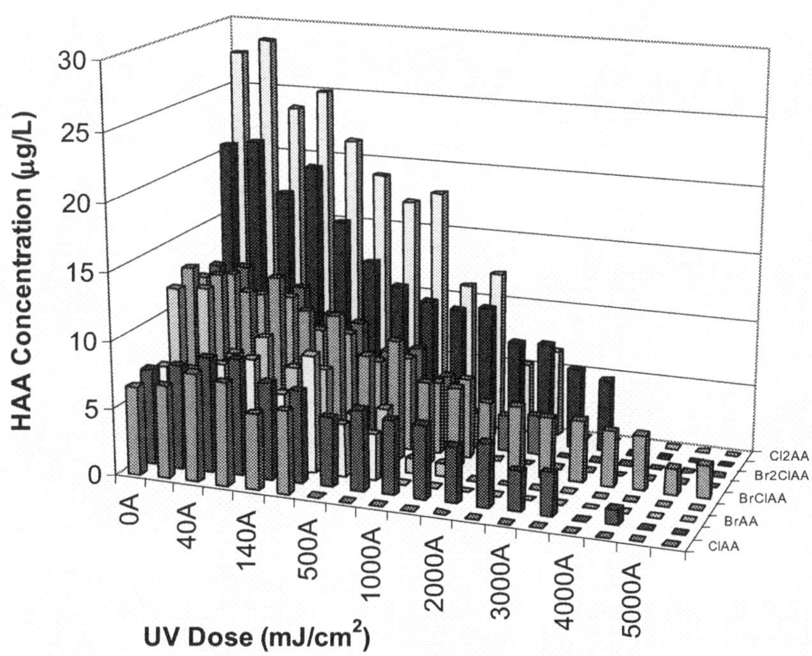

Figure 5.15 HAA concentrations from UV irradiation of synthetic water in the presence of nominally 128 µg/L NaBr and 100 mg/L H_2O_2, followed by UFC testing (showing all species detected, A/B replicate runs)

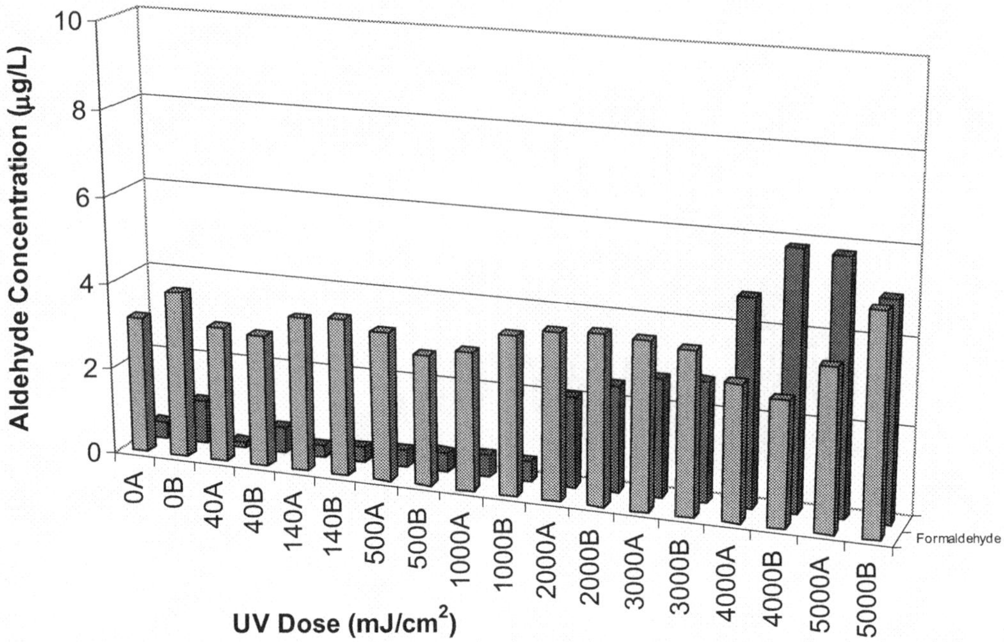

Figure 5.16 Aldehyde concentrations from UV irradiation of synthetic water containing 128 µg/L NaBr (showing all species detected, A/B replicate runs)

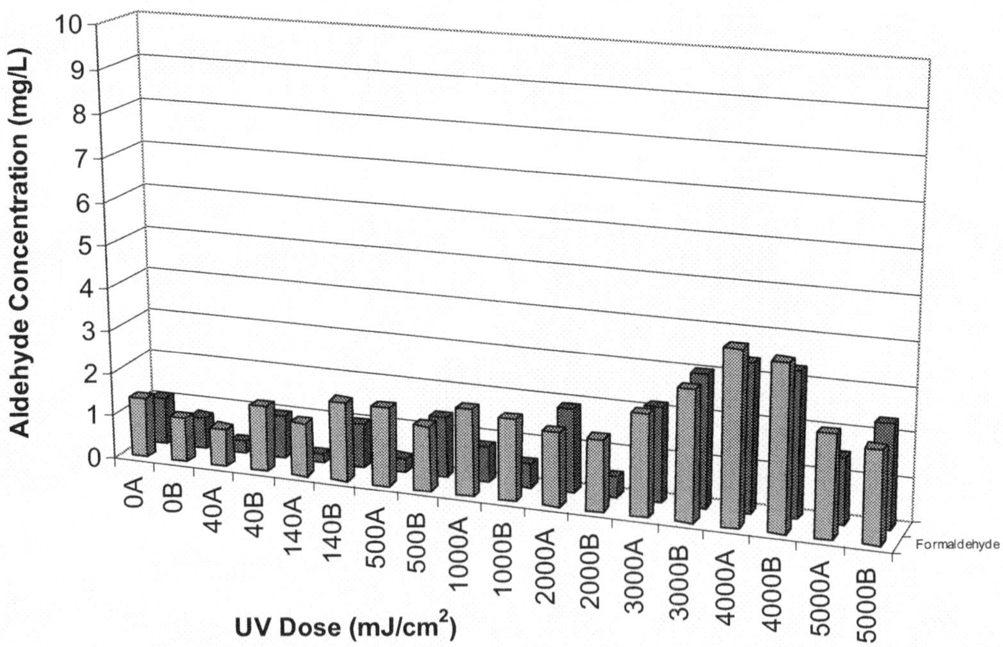

Figure 5.17 Aldehyde concentrations from UV irradiation of synthetic water in the presence of nominally 128 µg/L NaBr and 100 mg/L H_2O_2, (showing all species detected, A/B replicate runs)

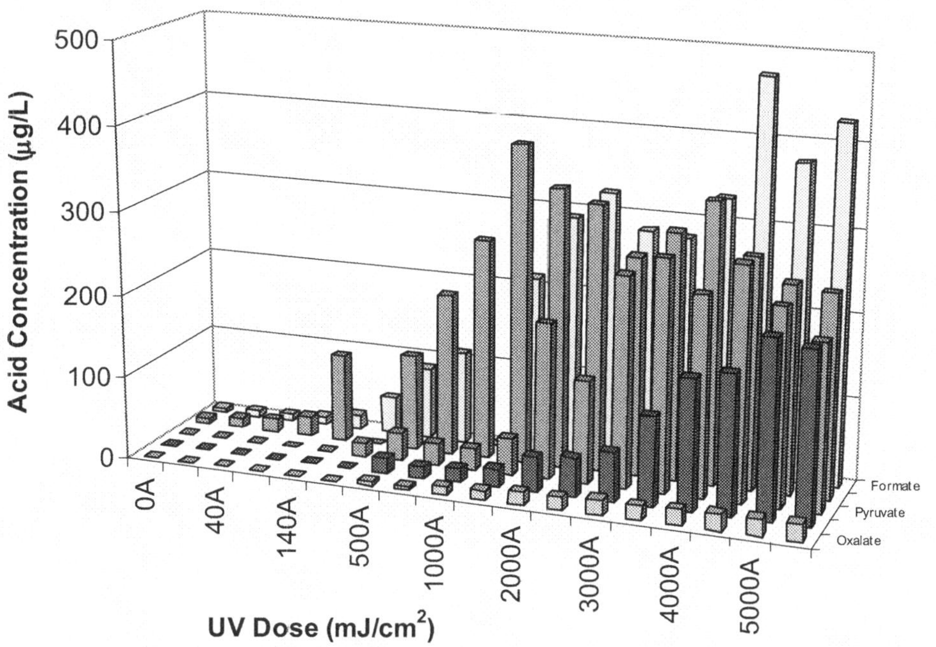

Figure 5.18 Carboxylic acid concentrations from UV irradiation of synthetic water containing 128 µg/L NaBr (showing all species detected, A/B replicate runs)

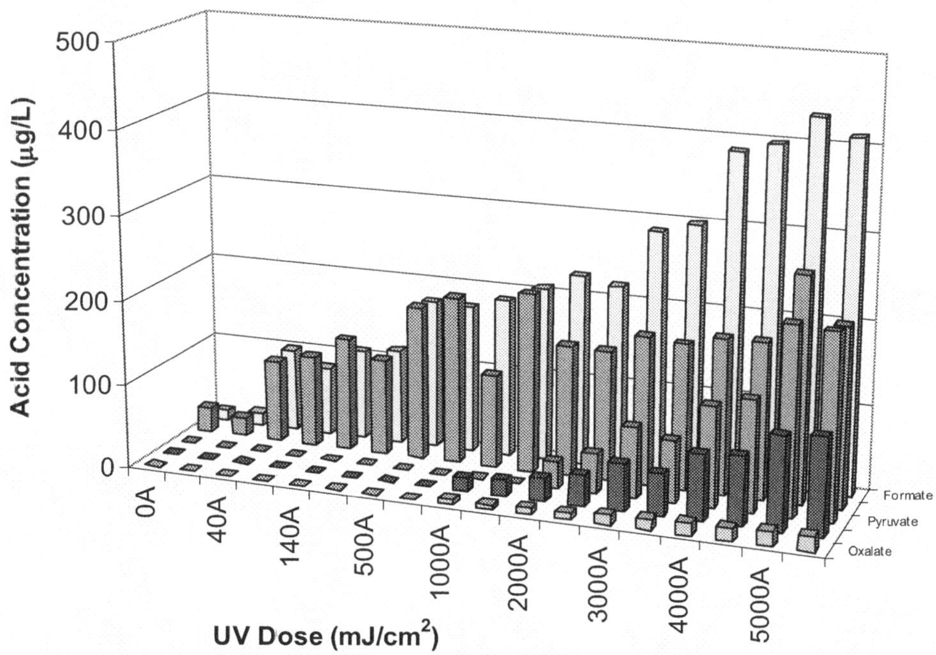

Figure 5.19 Carboxylic acid concentrations from UV irradiation of synthetic water in the presence of nominally 128 µg/L NaBr and 100 mg/L H_2O_2, (showing all species detected, A/B replicate runs)

CHAPTER 6
COMPARATIVE DEGRADATION OF CONTAMINANTS VIA UV PROCESSES

INTRODUCTION

The E_{EO} Basis for Comparing Susceptibility of Contaminants to Treatment

Advanced Oxidation Technologies (AOTs) are generally those that use electrically-driven processes to bring about the rapid oxidative degradation of organic pollutants in contaminated water or air. The oxidation reactions in these processes are essentially the same as the (very slow) degradative reactions that occur for these pollutants in the natural environment. The electrically-driven processes are perhaps a billion times faster than those in the natural environment, hence the term *Advanced* Oxidation Technologies. In some cases, the oxidation processes go to complete mineralization of the organic pollutants into CO_2, H_2O and inorganic acids for any S, F, N or Cl present.

It is important to be able to compare the electrical energy efficiency of various AOTs, since a large part of the operating cost is the cost of the electricity consumed. In 1998, the International Union of Pure and Applied Chemistry (IUPAC) recommended the figure-of-merit *Electrical Energy per Order* (E_{EO}) as a means of comparing the electrical energy efficiency for various AOTs. The E_{EO} is defined for a given AOT as *the number of kilowatt hours of electrical energy necessary to reduce the concentration of a given pollutant by one order of magnitude in a fixed volume of water* (e.g., 1,000 L or 1,000 U.S. gallons). The *smaller* the E_{EO}, the *more* efficient is the AOTs process.

The E_{EO} figure-of-merit is appropriate for any AOT process in which the degradation of a given pollutant follows first-order kinetics (almost always true for pollutant concentrations <10 mg/L). In such cases, it takes the same number of kWh of electrical energy to decrease the concentration from 1,000 ppb to 100 ppb as it does to go from 100 ppb to 10 ppb or from 10 ppb to 1 ppb. Thus, for example, in the UV/H_2O_2 treatment of a solution containing 1000 ppb of a given pollutant, it is found that the E_{EO} is 2.5 kWh/order/1,000 gal. If the cost of electricity is $0.08 per kWh and it is desired to reduce the concentration to 1 ppb (3 orders of magnitude), the electrical energy cost will be $3 \times 2.5 \times 0.08 = \$0.60$ per 1,000 gal treated.

Although the concept of Electrical Energy per Order is most often applied to ultraviolet-driven processes (e.g., direct UV photolysis, UV/ H_2O_2, UV/TiO_2, etc.), it is also applicable to any other AOT (e.g., electron beam irradiation or ozonation).

Mathematically, the E_{EO} can be defined for UV processes in a batch reactor in terms of the power consumption by the lamp and the time-based rate constant, k_t', according to equation 6.1:

$$E_{EO} = \frac{2.3 \times P \times 10^6}{V \times 60 \times k_t^{'}} \qquad 6.1$$

Here, P is the lamp power in W, V is the reactor volume in gallons, and k_t' is the rate constant in min^{-1}. With these units, the E_{EO} has units of kWh/1,000 gal, which is how they are expressed in this work. More detail concerning the E_{EO} approach and alternative formulations of equation 6.1 for flow-through systems are given elsewhere (Bolton and Stefan 2002).

EXPERIMENTAL APPROACH

Materials and Methods

The reactor and lamps used in this work are described in Chapter 2. The waters that were used were synthetic water 1 and filtered water from the Williams water treatment plant in Durham, NC. The MP lamp used in the reactor is different from the one used in the collimated beam experiments. Its emission spectrum is shown in Figure 6.1.

A stock solution of NDMA in water was used to spike the reactor, while for the mixed chemical contaminants the stock solution had to be prepared in dichloromethane (CH_2Cl_2) to allow dissolution of the compounds. An effort was made to ensure that the level of CH_2Cl_2 did not significantly compete with the NOM for ·OH, but practical constraints required a compromise. The concentration of the contaminants in the stock solution was 10,000 times higher than their initial concentration in the reactor. Thus, the concentration of CH_2Cl_2 in the reactor was 1.5 mM, and its contribution to ·OH scavenging was approximately 60% greater (1.6 times) than that of NOM in both synthetic and Durham water. Less of a compromise had to made for the mixed taste and odor compounds. For these, an aqueous stock solution was made from pure IPMP and methanolic solutions of MIB (10 mg/mL) and geosmin (2 mg/mL). These are the highest concentrations of MIB and geosmin that were available commercially, and their use introduces methanol as an ·OH scavenger during experiments at a concentration of 4.9 μM. At this level, methanol's contribution to the overall ·OH scavenging rate is approximately 65% that of the NOM in both synthetic and Durham water. Thus, for all experiments except those with NDMA, there is an organic solvent present that introduces some ·OH scavenging, but the overall scavenging rate is quite representative of those that can be expected in natural waters.

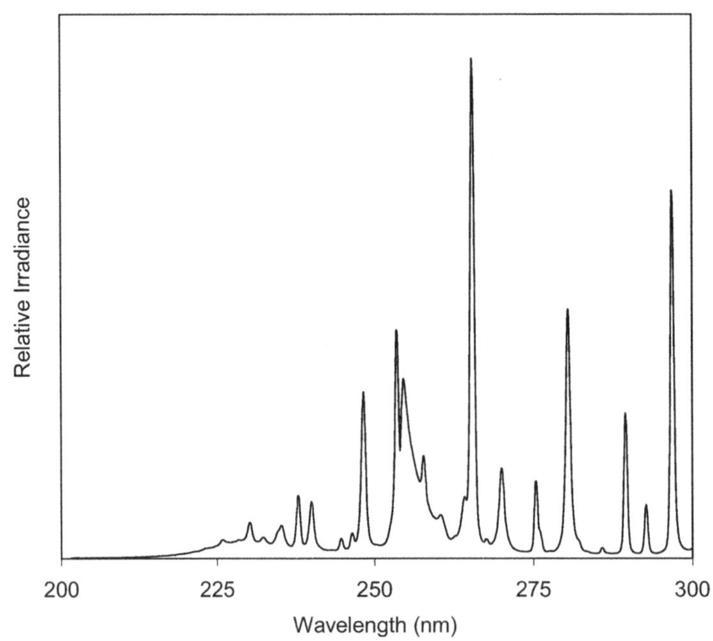

Figure 6.1 Emission spectrum of the MP lamp used in the bench-scale reactor

E_{EO} Measurement

To measure E_{EO}'s in a batch reactor with a known total volume, it is necessary only to quantify the time-based photolysis rate constant and the power drawn by the system. The time-based rate constants were quantified by first-order kinetic analysis of the removal rates as a function of time. Although the system in this case comprises a lamp, power supply, re-circulating water chiller, and mixer, only the power drawn by the power supply was used in E_{EO} values. The power was calculated from measurements of the voltage and current drawn from the electrical outlet into which the units were plugged. These measurements were taken simultaneously as diagramed in Figure 6.2 and the power was calculated as the product, (current x voltage). For the 500 W MP lamp, the current was sufficiently high (approximately 6 A) that it could be measured by a clamp-on type (inductance) ammeter, while for the 9 W LP lamp, the current was much lower (approximately 130 mA) and required measurement by connecting an ammeter in series with the circuit. Typical power requirements for the systems were 750 W and 15 W, for the MP and LP lamps respectively.

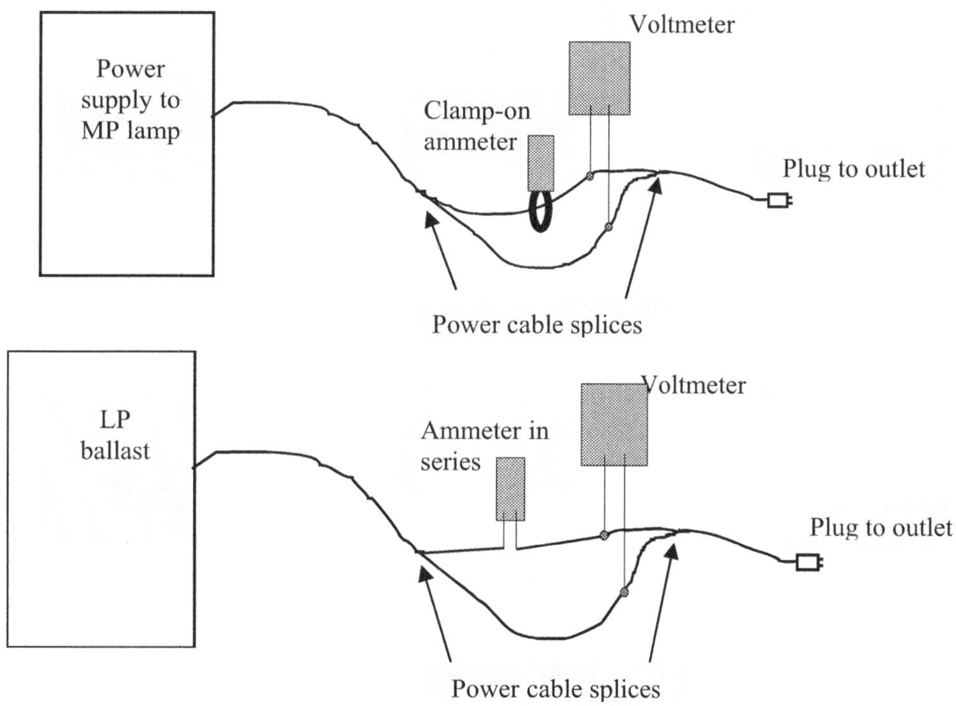

Figure 6.2 Experimental configurations for measuring power drawn by the MP and LP lamps in the bench-scale reactor

Experimental Protocol

To determine approximate dose rates in the reactor, the degradation of NDMA was followed, and appropriate exposure times for other contaminants were based on these measurements. The contaminants were split into three groups:

1. NDMA was run alone,
2. carbaryl, ATZ, toluene, metolachlor, and diazinon were run as a mixture, and
3. IB, geosmin, and IPMP were run as a mixture as well.

Initial contaminant concentrations were (in parentheses), NDMA (1 μM), carbaryl (0.3 μM), ATZ (0.6 μM), toluene (5 μM), metolachlor (1 μM), diazinon (2.5 μM). For the taste and odor compounds, the initial concentration was approximately 200 to 300 ng/L for all. All samples were run in dulpliates.

For MP experiments, the water in the reactor was initially cooled to 19 °C, and the lamp was then started. After five minutes of warming up, the lamp intensity was checked with a radiometer and any required chemicals were then spiked via syringe through the reagent addition port. The solution was allowed to mix for at least 10 minutes, at which point the temperature was approximately 25 °C. Samples were taken for initial concentration analysis and the experiment was commenced. For LP experiments, the protocol was the same except that the water was only cooled (or warmed) to 25 °C before commencing experiments. For chemical contaminants, four to six time points were sampled, while for taste and odor compounds, two or three time points were sampled.

RESULTS

Chemical Contaminants

As in collimated beam experiments, NDMA degradation was pseudo first-order under all conditions. The chemicals in the five-compound mixture also degraded by pseudo first-order kinetics with both LP and MP lamps as illustrated in Figure 6.3.

Using Equation 6.1, E_{EO} values were calculated and are presented graphically in Figure 6.4 for all the contaminants investigated. Note that the values are given as kWh/1,000 U.S. gal. Conversion to kWh/m^3 can be accomplished by multiplying by 0.26. Note that the y-axis in the figures goes only to 20 kWh/1,000 gal; both carbaryl and toluene are exceedingly slow to treat by UV alone, and for both waters and both lamps all E_{EO}'s except that for carbaryl with the MP lamp in Durham water are above 20.

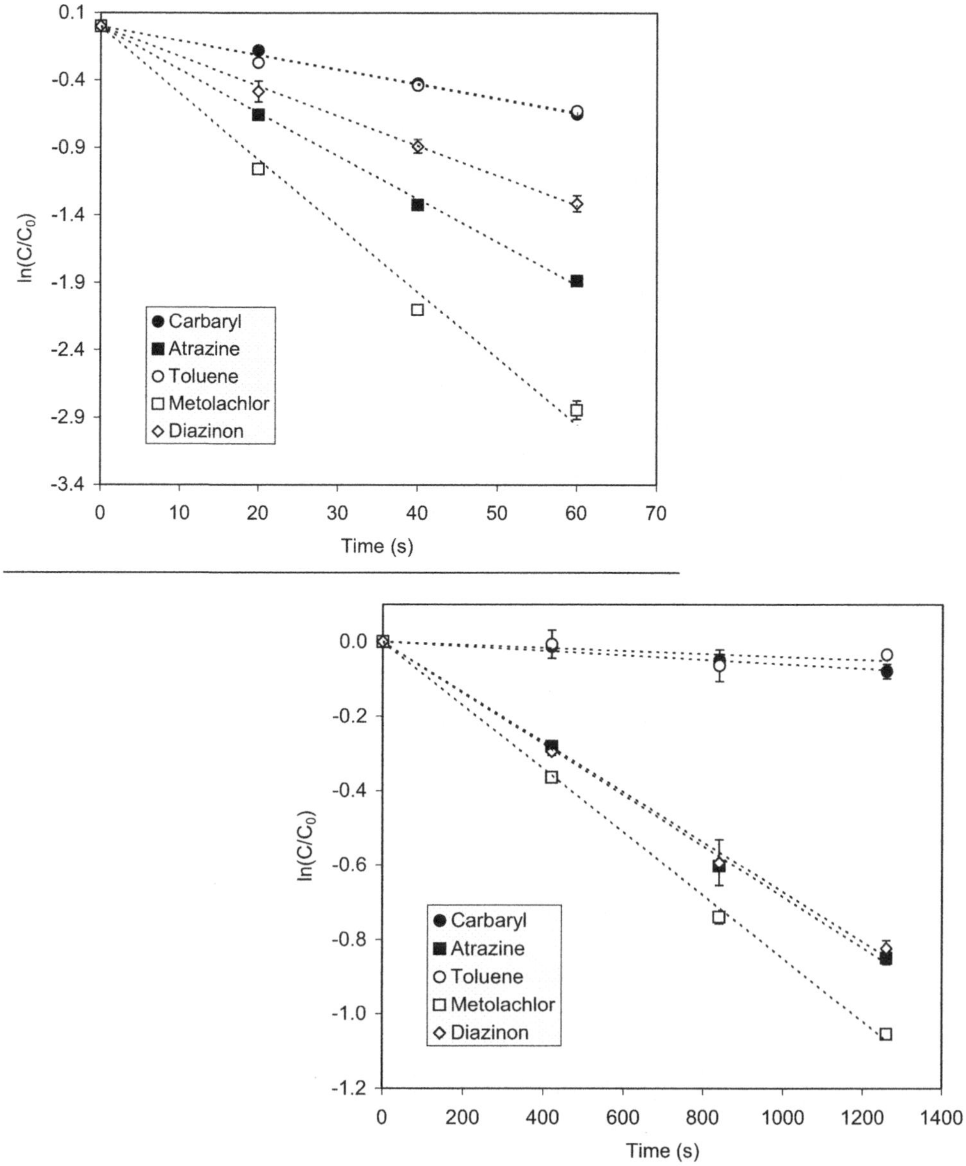

Figure 6.3 Photolysis by MP (top) and LP (bottom) lamps of five-compound mix in synthetic water 1 without H_2O_2

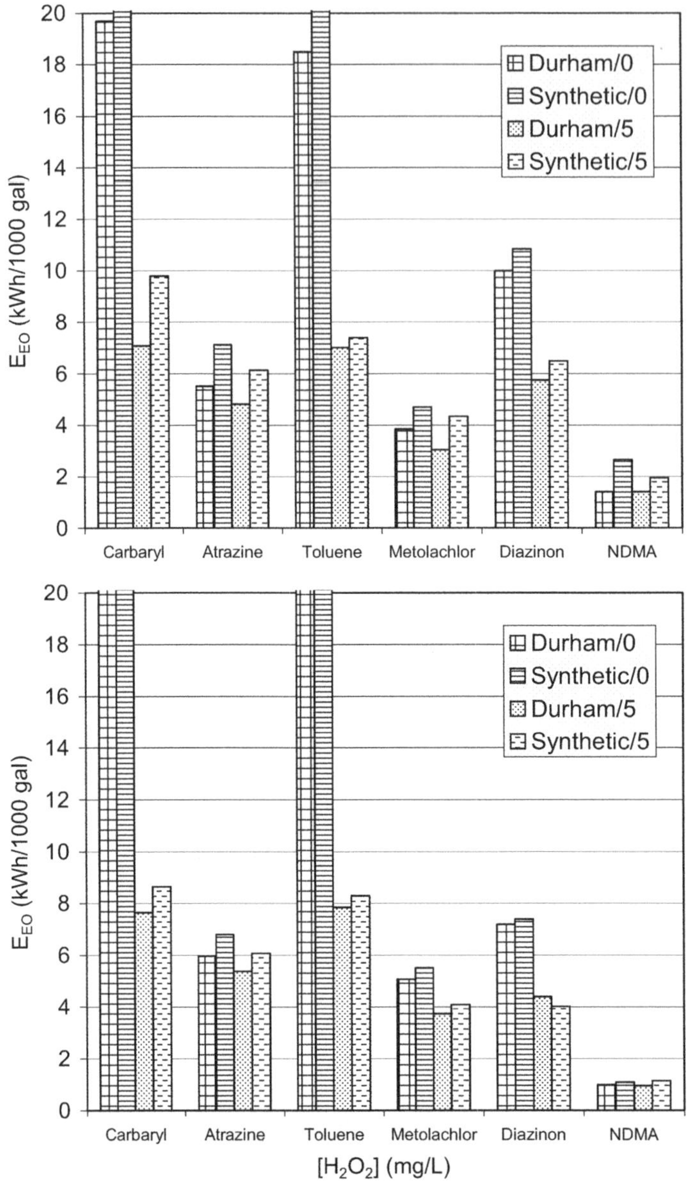

Figure 6.4 E_{EO} chart for MP (top) and LP (bottom) of all six contaminants tested under all conditions. Legend indicates type of water followed by mg/L of H_2O_2 applied

The observed trends are what would be expected: E_{EO}'s are lower in the Durham water, which has a higher transmittance and less NOM than the synthetic water, and they decrease with H_2O_2 addition for all compounds except for NDMA in the synthetic water with the MP lamp. This latter finding was unexpected and is presently unexplained. In Durham water, and with the LP lamp in synthetic water, H_2O_2 had almost no effect on NDMA degradation as predicted from collimated beam studies (Chapter 3).

In general, NDMA was the most treatable compound. The next most treatable compound was metolachlor, followed by atrazine and diazinon, all of which had appreciable rates of direct photolysis. Interestingly, the relative decrease in the E_{EO} with added H_2O_2 was greater for diazinon than ATZ, which suggests that it reacts more rapidly with ·OH than ATZ. However, a full analysis to obtain the actual rate constant is beyond the scope of the present discussion.

Carbaryl and toluene are both very unreactive with UV alone, and their E_{EO} values for direct photolysis are greater than 18 kWh/(1,000 gal) in all cases. With the addition of 5 mg/L H_2O_2, the E_{EO}'s drop substantially but still remain above 6.5 kWh/(1,000 gal) (1.7 kWh/m^3).

Taste and Odor Compounds

Due to limits on the number of taste and odor samples that could be analyzed, insufficient data were gathered to definitively conclude that the kinetics were first order, but the analysis assumes this is true due to the very low concentrations examined. Previous work (Chapter 4) indicated that direct photolysis of MIB was very slow. Therefore, most of the experiments performed here included H_2O_2 (either 2 or 5 mg/L). To verify previous results, however, one run was done without H_2O_2 using the MP lamp.

Using Equation 6.1, E_{EO} values were calculated and are presented graphically in Figure 6.5 for the taste and odor compounds. As with the chemical contaminants, the observed trends are expected: E_{EO}'s are lower in the Durham water and decrease with H_2O_2. The one exception to this trend is with the LP lamp and 5 mg/L H_2O_2, where MIB has a higher E_{EO} in Durham water than in synthetic water. This is quite possibly an artifact due to analytical precision and the small number of samples involved, because there is little reason to expect such a result based on all the other data. Also, as discussed below, the ratio of the E_{EO}'s for MIB and geosmin in the Durham water sample (LPD5) is quite different from all other runs and what is anticipated based on other researchers' data.

Due to analytical problems, the E_{EO} could not be calculated for geosmin using the MP lamp with 5 mg/L H_2O_2 in Durham water. However, based on the relative rates of reaction (and E_{EO}'s) of MIB and geosmin, one can estimate the expected value. Averaging the relative rates for all samples containing H_2O_2 except LPD5 (see above) shows that MIB reacts with $\cdot OH$ 1.4 ± 0.1 times as fast as geosmin. Based on this, and the E_{EO} for MIB in the same sample, a value of 0.92 kWh/(1,000 gal) is estimated for geosmin. Statistical analysis of the removal of the taste and odor compounds is presented in Appendix H.

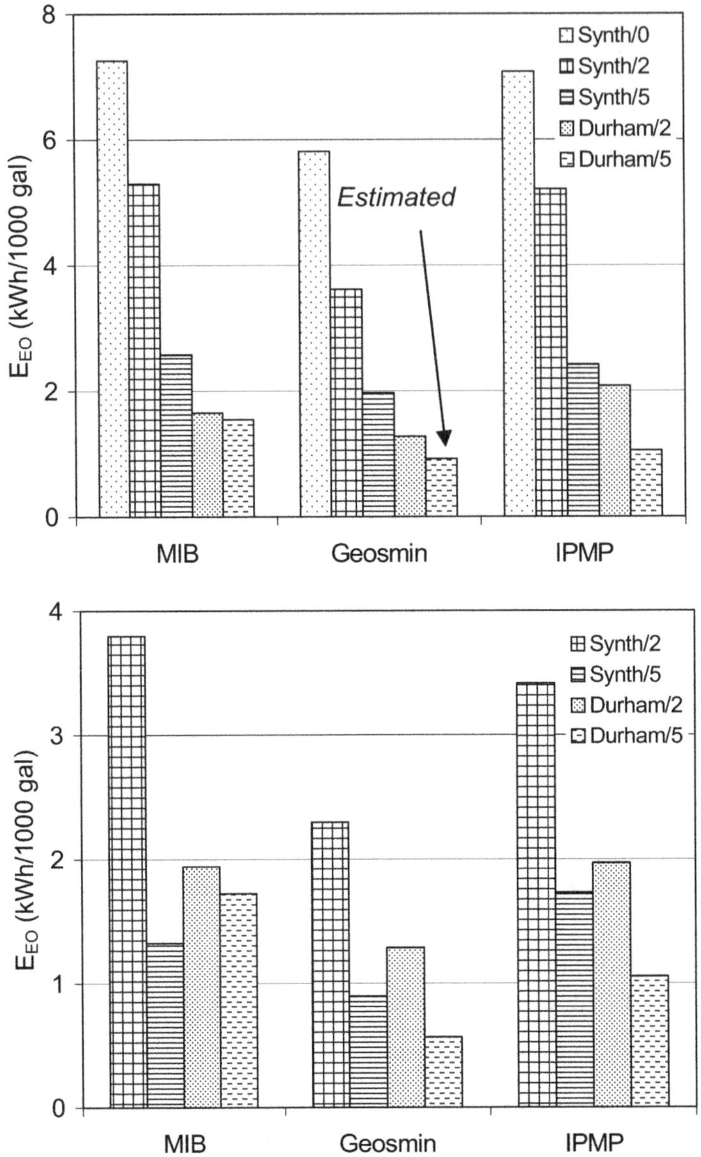

Figure 6.5 E_{EO} chart for MP (top) and LP (bottom) of taste and odor compounds for all conditions. Legend indicates type of water followed by mg/L of H_2O_2 applied. The value for geosmin in Durham water with 5 mg/L H_2O_2 and the MP lamp was estimated as described in the text

DISCUSSION

Bench-scale Versus Pilot-scale E_{EO} Values

In the following discussion, it must be recognized that the measured E_{EO}'s are highly dependent upon the reactor geometry and water quality and that they are not necessarily applicable to all reactors/waters. Nonetheless, the reactor used here was designed with attention to simulating as closely as possible the optical pathlength and ratio of lamp power to treated volume that might be found in a full-scale system. Thus, these E_{EO}'s are likely close to what can be expected in such systems. However, given the realities with flow-through hydraulic differences, these data are necessarily more qualitative than quantitative.

This is supported by the data for the chemical contaminants. For example, the E_{EO}'s for treating NDMA with the MP lamp can be compared with previously measured values in a pilot-scale system treating filtered re-use water. Bolton et al. found E_{EO}'s of between 0.5 and 1.4 kWh/(1,000 gal) for UV alone and 0.6 to 0.8 kWh/(1,000 gal) for UV/H_2O_2 with H_2O_2 levels of 5 to 10 mg/L (Bolton et al. 2001). In this work, E_{EO} values of 1.4 and 2.7 kWh/(1,000 gal) were found for UV alone in Durham and synthetic water, with the corresponding values being 1.4 and 1.9 kWh/(1,000 gal) for UV plus 5 mg/L H_2O_2. Given potential differences in lamp spectra, power consumption, and water quality that likely exist between this data and that of Bolton et al., the agreement is quite reasonable, within an order of magnitude.

Pilot scale E_{EO} values were also developed in the tests done at DWSD using both LP and MP reactors (see Chapter 7). Data were generated for atrazine, MIB and geosmin. The pilot scale data exhibited considerably lower E_{EO} values (see Table 7.4 and 7.5) for atrazine than the bench reactor results, however, the pilot electrical energies used for the calculation were the rated values and not the actual measured values, due to problems with the power meters. For example, atrazine E_{EO} values for the bench reactor were between 5 and 7 kWh/(1,000 gal) with and without peroxide for both LP and MP sources. For the pilot study the atrazine E_{EO} values were between 2 (with peroxide) and 3.5 (without peroxide) for the MP reactor and between 0.4 (with peroxide) and 2 (without peroxide) for the LP reactor. Again, the waters were also different in the 2 tests, suggesting that these numbers will be fairly site specific. MIB and geosmin E_{EO} values can also be compared and are much more similar. For LP UV the bench reactor values (with peroxide) were between 1 and 2 for MIB and 0.5 and 1.3 for geosmin. On the pilot scale, the E_{EO} values with peroxide were between 0.3 and 0.6 for MIB and 0.6 to 1 for geosmin. For MP UV the bench reactor values (with peroxide) were between 1.7 and 5 for MIB and 1 and 3.7 for geosmin. On the pilot scale, the E_{EO} values with peroxide were between 1.2 and 2.5 for MIB and 1.5 to 2.2 for geosmin.

Comparison Between LP and MP Lamps

Of interest to utilities considering UV for chemical oxidation are the relative efficiencies of LP and MP lamps. A very interesting finding in this work is that the relative efficiencies of LP and MP for treating NDMA and ATZ as judged by the E_{EO} values is deducible from the collimated beam studies in Chapter 3. In the discussion there it was hypothesized that LP would be more efficient than MP for treating NDMA while there would be little difference for ATZ. Indeed, for ATZ the difference between the MP and LP E_{EO} is less than 11% in all cases, but for NDMA it is approximately 30% lower in Durham water and 50% lower in synthetic water. These scenarios agree excellently with those predicted based on collimated beam studies alone (Chapter 3). Thus, careful consideration of spectral data and lamp power consumption can be combined with k_d' values measured in collimated beam instruments to make meaningful predictions of electrical efficiencies in actual reactors.

Of the other UV-susceptible contaminants, diazinon and metolachor, diazinon appears to be similar to NDMA in that the E_{EO} values are approximately 30% lower for the LP lamp. With metolachlor, the MP lamp offers a slight advantage of a 15-20% lower E_{EO} value compared to the LP lamp.

For treating the taste and odor compounds by UV/H_2O_2, the LP lamp was more efficient by a factor of approximately 1.4 to 2 in synthetic water depending upon the compound, while the lamps performed much more similarly in Durham water. Presumably, the higher light screening in the synthetic water at shorter wavelengths hinders the MP lamp performance.

The E_{EO} values can be transformed into approximate treatment costs by assuming a given electricity rate and treatment scenario. Table 6.1 presents the costs associated with treating the compounds by UV/H_2O_2 assuming an electricity rate of \$0.0815/kWh, the rate in Durham in July, 2003. All costs assume a 1-log reduction in concentration and that Durham filtered water is used along with 5 mg/L H_2O_2. Finally, all costs are for electricity alone and do not factor in the cost of H_2O_2 or system maintenance.

Table 6.1
Cost in \$U.S. to treat (1-log removal) 1000 gal of Durham water with UV/H_2O_2 (5 mg/L)

	ATZ	Carbaryl	Diazinon	Metolachlor	NDMA	Toluene	IPMP	Geosmin	MIB
MP	0.39	0.58	0.47	0.25	0.11	0.57	0.09	0.07	0.13
LP	0.44	0.62	0.36	0.30	0.08	0.64	0.09	0.05	0.14

Comparison Between the Different Chemicals

With the number of different compounds studied here, it is tempting to try to generalize about the photochemistry of individual compounds. However, the contaminants studied have widely varying chemical structures, and no general conclusions can be drawn about the basic nature of their photoreactivity without a broader study of structurally related compounds. Nonetheless, it is clear from the EEO values that of the contaminants examined here, ATZ, metolachlor, diazinon, and NDMA are far more photoreactive than are toluene and carbaryl. Of these chemicals, diazinon removal is most affected by H_2O_2 (Figure 6.4), suggesting that it has the highest kOH of the four photoreactive contaminants. For the taste and odor compounds, there is not so dramatic a difference between the E_{EO} values as with the contaminants, which is not surprising in the case of geosmin and MIB, which have fairly similar structures. However, it is somewhat surprising that IPMP with two nitrogen atoms possessing lone pair electrons does not undergo photolysis more rapidly than MIB and geosmin. The fact that H_2O_2 causes similar relative enhancements in the removal rates of the taste and odor compounds suggests that they all have similar k_{OH} values.

SUMMARY

Measurements of E_{EO} values in a carefully designed bench-scale reactor yield results that appear comparable to those expected in larger scale units. Interestingly, the costs associated with treatment of the taste and odor compounds IPMP, geosmin, and MIB with UV/H_2O_2 are very comparable to those for treating NDMA, both with LP and MP lamps. Of the remaining compounds, metolachlor is the most susceptible to UV/H_2O_2 treatment, and is more efficiently treated by LP/H_2O_2 than MP/H_2O_2. However, treating this compound will require approximately two to three times more electricity than NDMA for a given extent of removal. Much higher costs can be expected for treating ATZ and diazinon, and higher costs still for treating toluene and carbaryl.

CHAPTER 7
DEGRADATION OF CONTAMINANTS ON THE PILOT SCALE

INTRODUCTION

Ultraviolet (UV) radiation is increasingly being used as a primary disinfectant for drinking water. However, the ability of UV radiation alone or in combination with hydrogen peroxide or ozone (advanced oxidation processes) in oxidizing organic and organoleptic compounds has not been fully documented.

A pilot plant study was conducted using two UV treatment technologies: low-pressure high output (LPHO) and medium pressure (MP) UV lamps, using UV treatment alone and in combination with hydrogen peroxide, to evaluate the oxidation/degradation of two organoleptic chemicals (2-methylisoborneol, MIB, and geosmin) and the herbicide, atrazine (2-chloro 4-ethylamino 6-isopropylamino s-triazine). The pilot study was conducted at the City of Detroit's Southwest Water Treatment Plant (Southwest WTP), which is located in Allen Park, Michigan. The Southwest WTP finished water was used as feed water to the UV pilot plant.

PILOT PLANT FEED WATER QUALITY

The Southwest WTP is one of the five water treatment plants owned and operated by the City of Detroit Water and Sewerage Department (DWSD). The plant has a rated capacity of 160 millions gallons per day (mgd), and gets its raw water from the Detroit River (part of the Great Lakes system) through an intake off of Fighting Island. The treatment process consists of low-lift pumping; pre-chlorination, hydrofluosilicic acid addition (for fluoride), liquid alum addition (primary coagulant) and powdered activated carbon addition (used seasonally only for taste-and-odor event mitigation) in the rapid mix chamber; flocculation; sedimentation; dual-media filtration using anthracite/sand; followed by post-chlorination and phosphoric acid addition (for corrosion control); and high-lift pumping.

At the Southwest WTP, the pre-chlorination dose ranged from 1.5 to 2.0 mg/L during pilot testing. The hydrofluosilicic acid dose was 1 mg/L as fluoride. The alum dose varied from 0.7 to 2.0 mg/L as Al, with an average alum dose of 1.0 mg/L as Al. A post-chlorination dose of 0.1 to 0.3 mg/L was added to maintain a total residual chlorine (TRC) of 1.0 to 1.1 mg/L leaving the plant. The phosphoric acid dose was 1 mg/L as phosphate. The Southwest WTP raw water average turbidity is 3 to 7 ntu, with excursions up to 100 ntu during fall and spring. During pilot testing, the raw water turbidity ranged from 4 to 30 ntu. The raw water TOC was low and ranged from 1.5 to 2.5 mg/L, with an average of 1.8 mg/L.

The Southwest WTP finished water was used as feed water for the UV pilot plant. The Southwest WTP finished water is pumped into the distribution system using high lift pumps that pump into a two-tier header system. An 8-inch line off the header system provides service water within the plant. A 3-inch line that provides flushing water to the raw water screens (at approximately 70 psi pressure) was used as the feed water source to the UV pilot plant. A summary of the Southwest WTP finished water quality was presented in Table 2.3. The finished water has very low turbidity, low TOC, neutral to slightly basic pH, low to moderate alkalinity and hardness, and very low iron and manganese.

A transmittance scan was performed on the filtered water (prior to post-chlorination and phosphoric acid addition) and finished water samples, to determine if there was any absorbance interference in the wavelength range of 200 to 300 nm. The scan was performed using a 1-cm path length quartz cell. The results of this scan are presented in Figure 7.1.

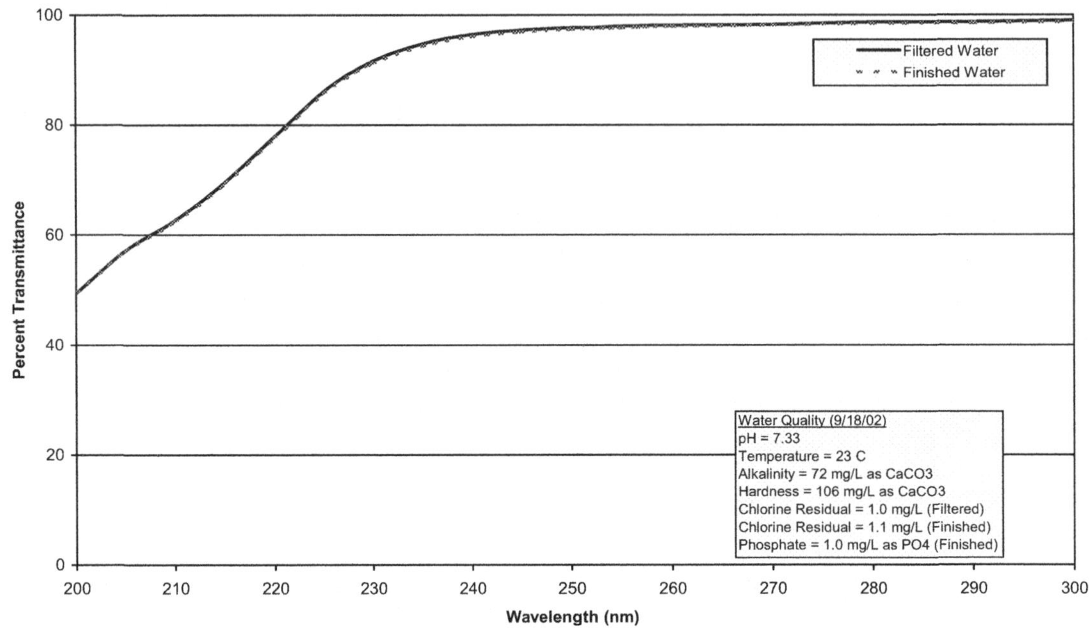

Figure 7.1 Absorbance scan of Southwest WTP filtered and finished water

Results from the two scans indicate that both the filtered and finished water samples had transmittance greater than 97 percent at a wavelength of 254 nm. The two scans were almost identical indicating that the addition of phosphoric acid and chlorine did not affect the transmittance in the 200-300 nm wavelength range.

EXPERIMENTAL SETUP

General Description of the UV Pilot Plant

The pilot plant consisted of two pilot UV reactor units – a low-pressure high-output lamp UV reactor unit (Trojan 8AL20 UV reactor, manufactured by Trojan Technologies, London, Canada) and a medium-pressure lamp UV reactor unit (Calgon 4 x 1 KW Sentinel reactor, manufactured by Calgon Carbon Corporation, Pittsburgh, U.S.A.). Figure 7.2 is a schematic of the UV pilot plant, while Figure 7.3 shows photographs of the UV pilot plant.

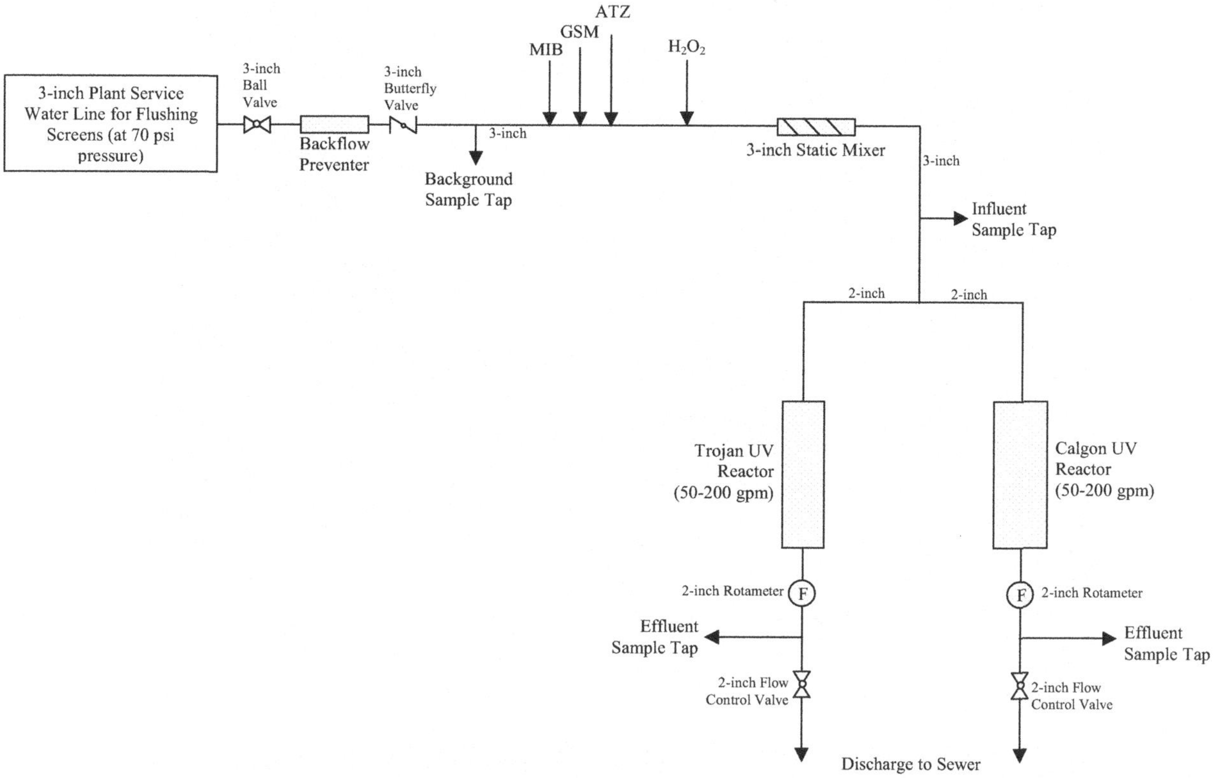

Figure 7.2 UV Pilot Plant Schematic

A 3-inch tee from the 3-inch Southwest WTP raw water screens flushing line provided feed water to the UV pilot plant. A 3-inch ball valve, a double-check backflow preventer, and a 3-inch butterfly valve were installed on the feed water line. With all valves fully open, the maximum flow through each UV reactor ranged from 190 to 200 gallons per minute (gpm).

A background sample tap was installed upstream of the chemical injection ports. Four ¼-inch injection ports with stopcocks were provided for injecting the chemicals (MIB, geosmin, atrazine, and hydrogen peroxide). During testing, however, MIB, geosmin and hydrogen peroxide were pumped from a single stock solution through one injection port; atrazine was pumped through another injection port. Due to high pressure in the feed water line (55 to 60 psi), Masterflex pumps with high performance pump heads (up to 100 psi) were used to meter the chemicals. A ¼-HP Masterflex pumps with 0-100 rpm speed and a high performance Masterflex peristaltic pump head was used to pump the MIB, geosmin and hydrogen peroxide mixture. This MIB, geosmin and hydrogen peroxide mixture was prepared daily. Due to the low solubility of atrazine (listed as 70 mg/L), a 50 mg/L atrazine stock solution was used. A ¼-HP Masterflex pump with 0-600 rpm speed and a Masterflex diaphragm pump head was used to pump the atrazine solution. The preparation of chemical stock solutions will be discussed in detail later.

Figures 7.3 Photographs of the UV pilot plant

A 3-inch, six-element in-line static mixer was installed on the feed water line downstream of the chemical injection ports to uniformly mix the chemicals. A sample tap was provided downstream of the static mixer to collect influent samples (common to both UV reactors). The flow was then split into two 2-inch parallel lines to feed the two UV reactors, with a ball valve upstream of each reactor to isolate that particular reactor. A 2-inch rotameter (calibrated between 50 and 200 gpm flow), an effluent sample tap, and a control ball valve were installed downstream of each UV reactor. The effluent from both reactors was discharged to the local sewer.

Description of the Two UV Reactors

The LPHO UV unit was Trojan Technologies' 8AL20 UV reactor. The LPHO UV reactor consisted of eight (8) LPHO UV lamps located axially in an 8-inch diameter chamber. The 4-inch inlet and outlet flanges were reduced to 2-inch flanged ends. The power supply consisted of 2.25 KW, 208-240 V, 60 Hz, single phase, three-wire system. The LPHO UV reactor was rated for flows between 50 and 250 gpm, with a maximum head loss of 6 inches of water. A control panel was provided for powering the UV reactor, and to monitor UV lamp status, lamp intensity and hours of lamp service. Figure 7.4 is a layout of the LPHO UV reactor.

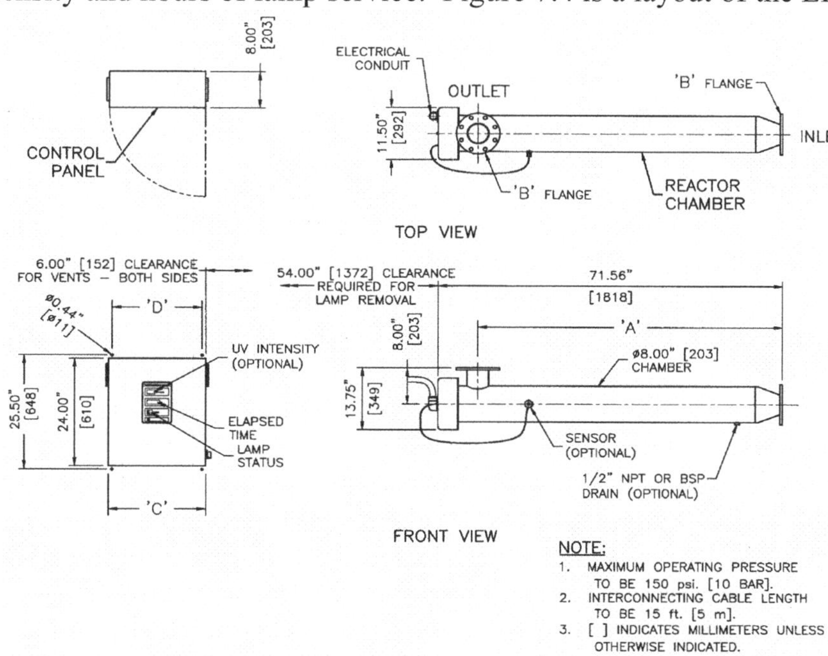

Figure 7.4 Schematic of the LPHO UV reactor (Courtesy of Trojan Technologies, Inc.)

The MP UV reactor unit was Calgon Carbon Corporation's 4 x 1 KW Sentinel reactor. The MP UV reactor consisted of a 12-inch diameter chamber with the 4 MP UV lamps installed at 90-degrees to the direction of flow. The 12-inch inlet and outlet flanges were reduced to 2-inch flanged openings. The reactor is designed for a maximum water pressure of 100 psi. The recommended rate of flow through the MP UV reactor is 50 to 700 gpm, with a maximum head loss of 5 psi at 700 gpm flow. The electrical system required 120 VAC for the PLC and logic components. The reactor was also equipped with UV intensity sensors. The control panel was used to power the UV reactor, and to monitor UV lamp intensities and hours of operation. Figure 7.5 is a schematic of the MP UV reactor.

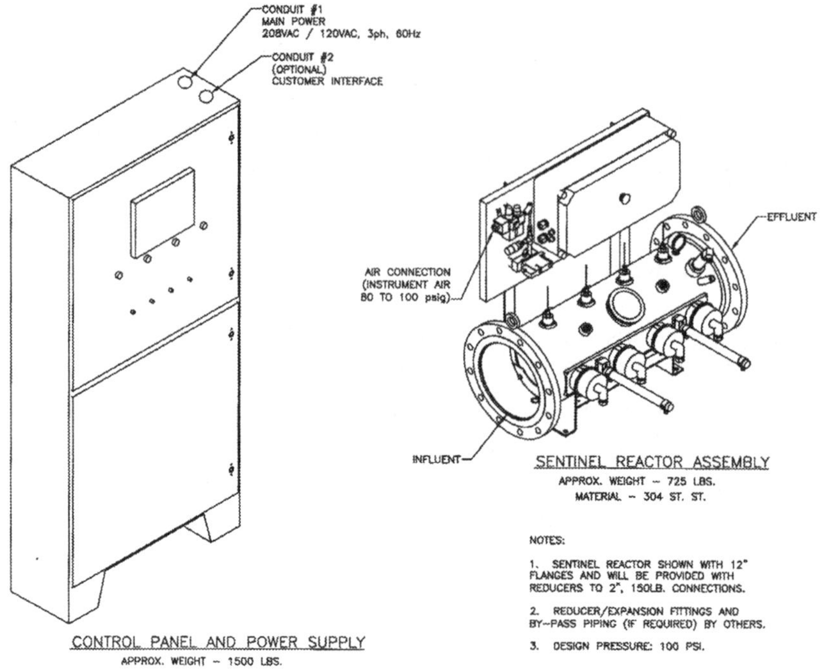

Figure 7.5 Schematic of the MP UV reactor (Courtesy of Calgon Carbon Corp.)

TESTING PROTOCOL

Two sets of pilot experiments were conducted during the course of this project. The first set of experiments was conducted in October and November 2002. Due to problems with sample preservation, the tests were repeated in April 2003. Only the April 2003 testing protocol and results are discussed in this report as the 2002 data were deemed unreliable.

The April 2003 pilot tests consisted of twenty-eight (28) tests. Table 7.1 is a description of test conditions for these 28 pilot tests.

The test matrix consisted of three hydrogen peroxide doses (0, 4 and 6 mg/L), three flow conditions (50, 100 and 190 gpm), and two levels of MIB, geosmin and atrazine concentrations (low level of 150 ng/L, 150 ng/L and 20 μg/L; and high level of 300 ng/L, 300 ng/L and 40 μg/L, respectively). To achieve the maximum flow of 190 gpm through one UV reactor, the flow through the other UV reactor had to be bypassed. This resulted in a total of 24 tests. In addition, test 1 was run as a control where no chemicals were added. Also, tests 26 through 28 were run at conditions identical to test 14. Hence, a total of 28 tests were conducted.

Atrazine was purchased from Sigma-Aldrich Company. The atrazine stock solution was prepared in 5-gallon containers by dissolving 0.95 grams of atrazine in 5 gallons of water (50 mg/L stock concentration). Since atrazine dissolution is slow, the atrazine stock solution was mixed using a magnetic stirrer for a minimum of two days. After that, the solution was filtered using a 0.45 μm pore size filter to remove any particulate matter.

Table 7.1
UV pilot plant test protocol

Test No.	Flow (gpm) LPHO UV Reactor	Flow (gpm) MP UV Reactor	H_2O_2 (mg/L)	MIB (ng/L)	Geosmin (ng/L)	Atrazine (µg/L)
1 (Control)	50	50	0	0	0	0
2	50	50	0	150	150	20
3	100	100	0	150	150	20
4	190	-	0	150	150	20
5	-	190	0	150	150	20
6	50	50	0	300	300	40
7	100	100	0	300	300	40
8	190	-	0	300	300	40
9	-	190	0	300	300	40
10	50	50	4	150	150	20
11	100	100	4	150	150	20
12	190	-	4	150	150	20
13	-	190	4	150	150	20
14*	50	50	4	300	300	40
15	100	100	4	300	300	40
16	190	-	4	300	300	40
17	-	190	4	300	300	40
18	50	50	6	150	150	20
19	100	100	6	150	150	20
20	190	-	6	150	150	20
21	-	190	6	150	150	20
22	50	50	6	300	300	40
23	100	100	6	300	300	40
24	190	-	6	300	300	40
25	-	190	6	300	300	40
26*	50	50	4	300	300	40
27*	50	50	4	300	300	40
28*	50	50	4	300	300	40

* Tests 26, 27 and 28 were identical to Test 14

Fifty mg each of pure MIB and GSM (as powder) were purchased from Dalton Chemical Laboratories (Toronto, Canada) and Aroma and Fine Chemicals (UK), respectively. This MIB and GSM were dissolved in 5 mL of methanol and was used as the primary MIB-GSM stock solution (10 mg/mL each of MIB and GSM). The concentration of a 35-percent strength hydrogen peroxide solution (expected to be 395 g/L) was measured twice before the pilot study and was found to be 380 g/L. This measured value was used to make secondary hydrogen peroxide stocks. The primary MIB-GSM stock solution and the 35 percent hydrogen peroxide solution were used to prepare the daily MIB-GSM-H_2O_2 stock solutions (discarded at the end of the day).

For example, for tests 10 through 13, 0.2 mL of the 10 mg/mL primary MIB-GSM stock solution and 140 mL of the 380 g/L H_2O_2 stock solution were made up to a 500 mL volume using deionized water. This resulted in a stock solution with MIB and GSM concentrations of 4 mg/L and a H_2O_2 concentration of 106.6 g/L. For test 10, for a total influent flow of 100 gpm, a pumping rate of 14.2 mL/min resulted in MIB and GSM concentrations of 150 ng/L and a H_2O_2 concentration of 4 mg/L. The 50 mg/L atrazine stock solution was pumped at 151 mL/min for a resulting atrazine concentration of 20 μg/L.

The chemical doses listed in Table 7.1 were the intended doses. However, the actual influent chemical doses may be slightly different due to limitations in the accuracy of the metering pumps, and mixing by the in-line static mixer. Therefore, for each test, influent samples were collected for MIB, GSM, atrazine and TOC analyses. Similarly, effluent samples from both UV reactors were also collected for the same analyses. These samples were stored in a refrigerator at 4°C and shipped to the corresponding laboratory for analyses. In addition, for tests where hydrogen peroxide was added, influent and effluent samples were collected and analyzed immediately for the H_2O_2 residual. H_2O_2 was measured in the field using the tri-iodide method. For a 2.5 mL aliquot that was made up to 10 mL, the H_2O_2 concentration (in mg/L) was equal to:

$$= (\text{Abs}_{351\ \text{nm}}, \text{ in cm}^{-1} / 26{,}450\ \text{M}^{-1}\text{cm}^{-1}) * (10 / 2.5) * 34{,}000\ \text{mg/mole}$$

$$\text{or, } H_2O_2, \text{ in mg/L} = (\text{Abs}_{351\ \text{nm}}, \text{ in cm}^{-1})*(5.142)$$

Samples were collected in 250-mL amber glass bottles with teflon-coated caps. The samples were collected headspace free. During tests 1 through 9 where no H_2O_2 was added, 0.5 mL of a 1,000 mg/L sodium sulfite solution was added to each sample bottle to quench the chlorine residual. During tests 10 through 28 when H_2O_2 was added, the residual H_2O_2 was expected to quench the chlorine and sodium sulfite was not added. Instead, 0.5 mL of a 100 mg/L bovine liver catalase solution (purchased from Sigma-Aldrich Corporation) was added to each sample bottle to quench the H_2O_2 residual.

RESULTS AND DISCUSSION

The results from the 28 pilot tests are presented in Table 7.2. Appendix H presents a statistical analysis of the MIB and GSM results for the pilot tests.

Table 7.2
UV pilot study results

Test No.	H_2O_2 Residual, mg/L			MIB, ng/L			GSM, ng/L			Atrazine, μg/L		
	Infl	LPHOUV Effl	MPUV Effl	Infl	LPHOUV Effl	MPUV Effl	Infl	LPHOUV Effl	MPUV Effl	Infl	LPHOUV Effl	MPUV Effl
1	-	-	-	BDL	BDL	BDL	BDL	BDL	BDL	3.1	0.3	0.3
2	-	-	-	122.3	85.8	73.8	108.7	62.6	61.5	13.5	5.0	5.1
3	-	-	-	142.5	107.2	113.3	129.9	101.6	78.0	14.2	8.4	7.4
4	-	-	-	150.1	116.0	-	117.0	109.9	-	14.7	10.2	-
5	-	-	-	166.0	-	115.2	125.1	-	101.9	13.0	-	10.1
6	-	-	-	278.1	153.3	145.5	221.9	143.0	109.1	25.5	11.0	10.2
7	-	-	-	293.0	229.3	195.1	258.5	175.2	154.0	31.5	18.8	18.0
8	-	-	-	265.5	235.0	-	230.0	194.0	-	28.8	21.6	-
9	-	-	-	280.8	-	217.1	225.2	-	184.7	28.8	-	22.6
10	3.0	2.9	3.0	141.1	10.5	28.1	117.0	10.4	21.2	13.1	2.2	3.5
11	3.3	3.3	3.3	144.8	42.9	53.9	132.9	28.7	37.6	13.8	5.6	6.9
12	3.0	3.1	-	136.2	76.1	-	108.9	64.2	-	12.7	8.3	-
13	3.3	-	3.3	139.2	-	85.7	121.0	-	66.0	14.1	-	9.4
14	3.1	3.1	3.2	257.4	30.5	58.6	215.8	13.5	43.5	24.2	5.3	7.1
15	3.4	3.5	3.5	303.6	85.5	115.2	255.3	57.5	70.5	27.3	11.6	13.9
16	3.2	3.2	-	254.8	155.7	-	205.9	114.9	-	26.1	17.1	-
17	3.4	-	3.5	271.8	-	167.2	212.7	-	118.1	28.3	-	18.2
18	5.0	4.8	5.0	148.7	1.0	9.4	109.7	7.1	7.8	11.3	0.3	1.9
19	5.6	5.6	5.7	139.6	13.6	24.9	140.0	8.5	23.0	12.5	3.1	4.3
20	5.4	5.4	-	146.0	50.7	-	124.8	35.4	-	11.6	6.1	-
21	5.8	-	5.8	149.2	-	62.0	133.1	-	49.6	13.0	-	6.7
22	5.2	5.0	5.1	284.1	6.0	44.2	219.5	13.4	21.1	24.0	2.1	5.3
23	5.7	5.7	5.7	296.7	39.5	77.5	256.3	19.6	46.9	28.1	6.0	9.9
24	5.4	5.4	-	274.7	107.5	-	249.4	74.1	-	25.9	13.5	-
25	5.7	-	5.8	298.3	-	129.9	236.6	-	84.4	28.0	-	14.8
26	NM	NM	NM	242.3	13.0	46.8	198.4	24.4	35.2	25.1	4.1	5.9
27	NM	NM	NM	227.9	18.8	48.9	191.6	11.3	32.5	24.2	3.9	7.2
28	NM	NM	NM	242.1	13.0	40.7	200.0	16.6	28.5	25.6	4.3	7.0

BDL – Below detection limit (1.0 ng/L for MIB and GSM; 0.2 μg/L for atrazine); NM – Not measured

For tests with a 4 mg/L H_2O_2 dose, the H_2O_2 residual varied from 2.9 to 3.5 mg/L. For tests with a 6 mg/L H_2O_2 dose, the H_2O_2 residual varied from 4.8 to 5.8 mg/L.

The percent degradation of MIB, GSM and atrazine were calculated for the LPHO and the MP UV reactors for the 28 tests. These results are summarized in Table 7.3.

Table 7.3
MIB, geosmin and atrazine percent degradation

Flow, gpm	H_2O_2, mg/L	Low or High Chemical Doses	Percent Degradation					
			MIB		GSM		Atrazine	
			LPHO UV	MP UV	LPHO UV	MP UV	LPHO UV	MP UV
50	0	Low	29.9	39.7	42.4	43.4	62.8	62.3
100	0	Low	24.8	20.5	21.8	40.0	40.4	47.8
190	0	Low	22.7	30.6	6.0	18.5	30.8	22.3
50	0	High	44.9	47.7	35.6	50.9	56.8	60.1
100	0	High	21.8	33.4	32.2	40.4	40.3	43.0
190	0	High	11.5	22.7	15.7	18.0	25.0	21.4
50	4	Low	92.5	80.1	91.1	81.9	83.4	72.9
100	4	Low	70.4	62.8	78.4	71.7	59.7	49.7
190	4	Low	44.1	38.5	41.0	45.5	34.3	33.4
50	4	High	88.1	77.2	93.7	79.8	78.1	70.8
			94.6	80.7	87.7	82.2	83.8	76.5
			91.8	78.6	94.1	83.0	83.9	70.4
			94.6	83.2	91.7	85.8	83.2	72.5
			Avg =92.3	***79.9***	***91.8***	***82.7***	***82.3***	***72.5***
			S.D.=3.1	***2.6***	***2.9***	***2.4***	***2.8***	***2.8***
100	4	High	71.9	62.1	77.5	72.4	57.5	49.0
190	4	High	38.9	38.5	44.2	44.5	34.6	35.7
50	6	Low	99.3	93.7	93.6	92.9	97.8	83.3
100	6	Low	90.2	82.2	94.0	83.5	74.8	65.4
190	6	Low	65.2	58.5	71.6	62.7	47.6	48.5
50	6	High	97.9	84.4	93.9	90.4	91.4	78.0
100	6	High	86.7	73.9	92.4	81.7	78.8	64.9
190	6	High	60.9	56.4	70.3	64.3	47.9	47.0

NOTES: Bold, italicized numbers are average and standard deviation values for the four tests
"Low" refers to MIB/GSM/atrazine doses of 150 ng/L/150 ng/L/20 μg/L
"High" refers to MIB/GSM/atrazine doses of 300 ng/L/300 ng/L/40 μg/L

Tests 14 and 26-28 were conducted at identical conditions. The individual as well as average (in bold, italicized) percent degradation for this condition are presented in Table 7.3. There was very low variability in the results for these four tests, which indicated that the test conditions were well controlled and the results were reproducible. The average values for this condition will be used for comparison purposes.

The objective of this pilot study was not to compare the performance of LPHO UV reactors versus MP UV reactors; but to document the degradation of organoleptic (MIB and GSM) and organic (atrazine) chemicals by the two UV technologies.

Electrical Energy per Order, E_{EO}

The parameter E_{EO} is a commonly used metric to evaluate the efficiency of UV reactors, when used for chemical contaminant treatment. The E_{EO} is defined as the electrical energy dose

(EED) required to reduce the contaminant concentration by one order of magnitude (one-log or 90 percent removal). The EED is defined as the measured reactor power draw divided by the reactor flow, and is expressed in units of kW-hr/1000gal or kW-hr/m^3.

Typical units used for E_{EO} are kW-hr/1000gal/order or kW-hr/m^3/order. The E_{EO} value is specific to the UV reactor, chemical contaminant, and water quality. Hence, different contaminants may have different E_{EO} values for the same UV reactor and feed water quality. The E_{EO} is a lumped metric that is a measure of the UV reactor's hydraulic, optical and electrical efficiencies. The E_{EO} value is directly proportional to the electrical draw (i.e., a lower E_{EO} value implies a lower electrical draw and vice versa).

The formula for computing E_{EO} is:

$$E_{EO} = EED / \log(C_0/C)$$

or, $E_{EO} = E / (Q*0.06*\log(C_0/C))$

where, EED is the electrical energy dose, in kW-hr/1000gal;
C_0 and C are the reactor influent and effluent concentrations, respectively;
E is the measured reactor power draw, in kW;
and, Q is the reactor flow rate, in gpm.

In this pilot study, the reactor power draws were not measured during the pilot tests. Hence, the rated electrical draws (2 kW for the LPHO UV reactor, and 4 kW for the MP UV reactor) were used to calculate the E_{EO} values. All E_{EO} values in this report are expressed as kW-hr/1000gal/order.

The correlation between reactor flows and the electrical energy dose (EED) for the LPHO and MP UV reactors is presented below in Table 7.4.

Table 7.4
Correlation between reactor flow rates and EED

Reactor Flow Rate (gpm)	Electrical Energy Dose, EED (kW-hr/1000gal)	
	LPHO UV Reactor	MP UV Reactor
50	0.67	1.33
100	0.33	0.67
190	0.18	0.35

LPHO UV Reactor

Figures 7.6 through 7.8 are plots of percent degradation of MIB, GSM and atrazine versus EED, for the LPHO UV reactor. Each figure shows the percent degradation of the chemical (MIB, GSM or atrazine) plotted against the electrical energy dose for six conditions – three hydrogen peroxide doses (0, 4 and 6 mg/L); at low and high levels of MIB, GSM and atrazine (150 ng/L, 150 ng/L and 20 μg/L; and 300 ng/L, 300 ng/L and 40 μg/L, respectively).

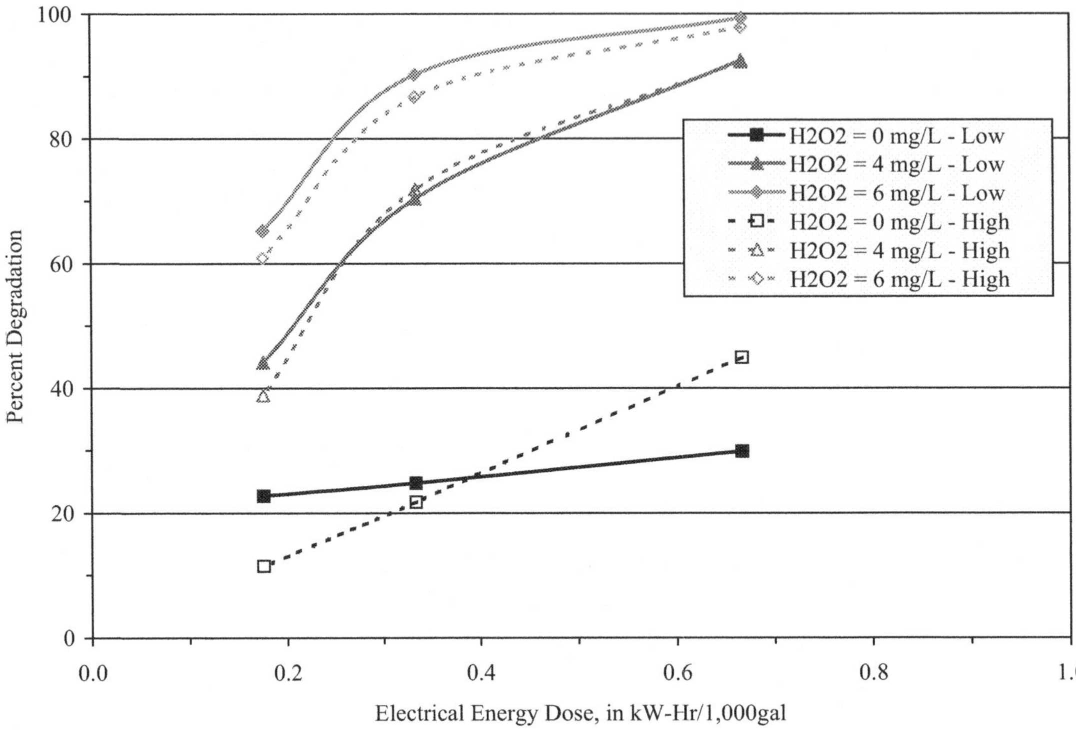

Figure 7.6 Percent degradation of MIB vs. EED for the LPHO UV reactor

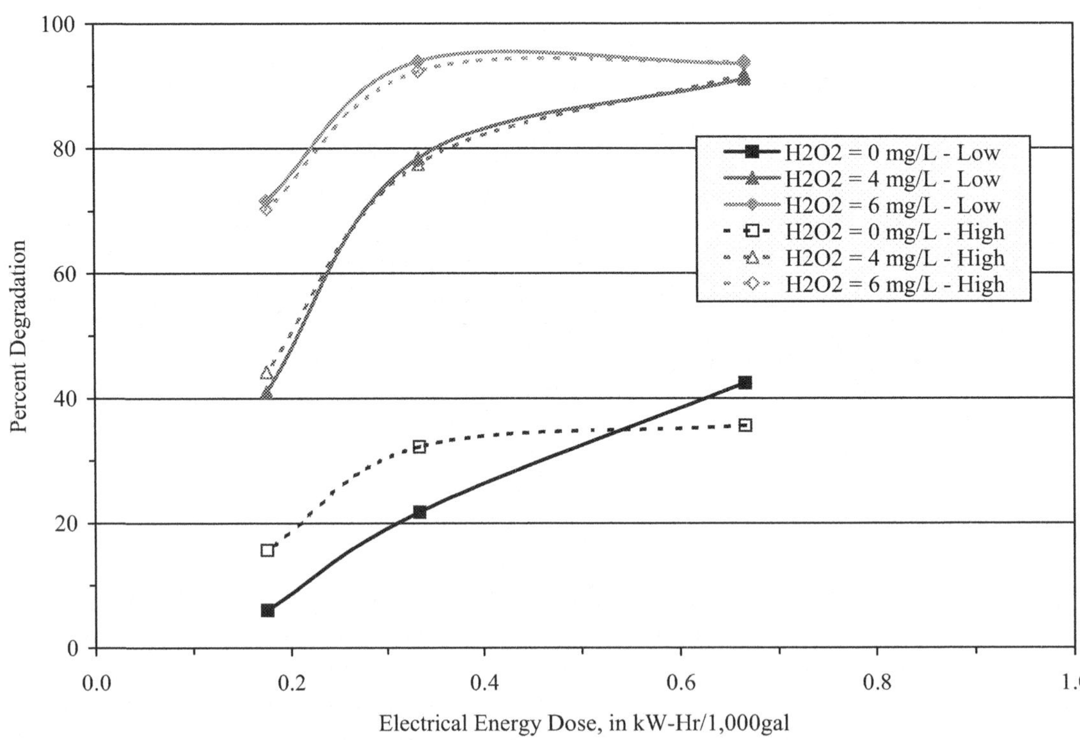

Figure 7.7 Percent degradation of geosmin vs. EED for the LPHO UV reactor

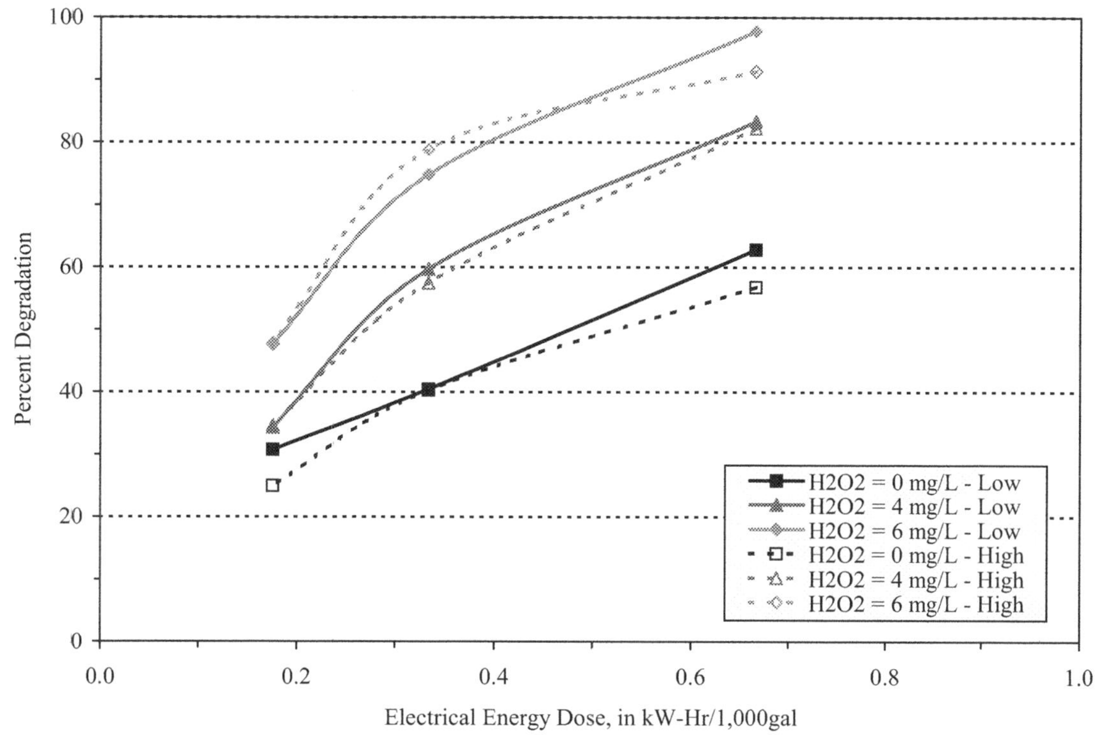

Figure 7.8 Percent degradation of atrazine vs. EED for the LPHO UV reactor

For the LPHO UV reactor, the degradation of MIB, GSM and atrazine was directly proportional to the EED (or inversely proportional to the flow), and directly proportional to the H_2O_2 dose. In general, the concentration of the chemical did not affect the percent degradation. However, when H_2O_2 was not being added, there appeared to be a greater variation in percent degradation between the low and high levels.

For MIB and GSM, the percent degradation with no H_2O_2 added was low (10 to 45 percent). With the addition of H_2O_2, there was a drastic increase in MIB and GSM degradation (rising to 40 to 99 percent).

For atrazine, there was significant degradation without the addition of H_2O_2 (30 to 60 percent). The percent degradation increased with increasing H_2O_2 dose, but the effect was more gradual (rising to 50 to 98 percent).

Figures 7.9 through 7.11 are graphs of $\log_{10}$ (C_{inf}/C_{eff}) plotted against the EED for MIB, GSM and atrazine. A linear regression was fit for each of the six data sets, and the E_{EO}'s were calculated for each data set. The E_{EO}'s are tabulated in Table 7.5.

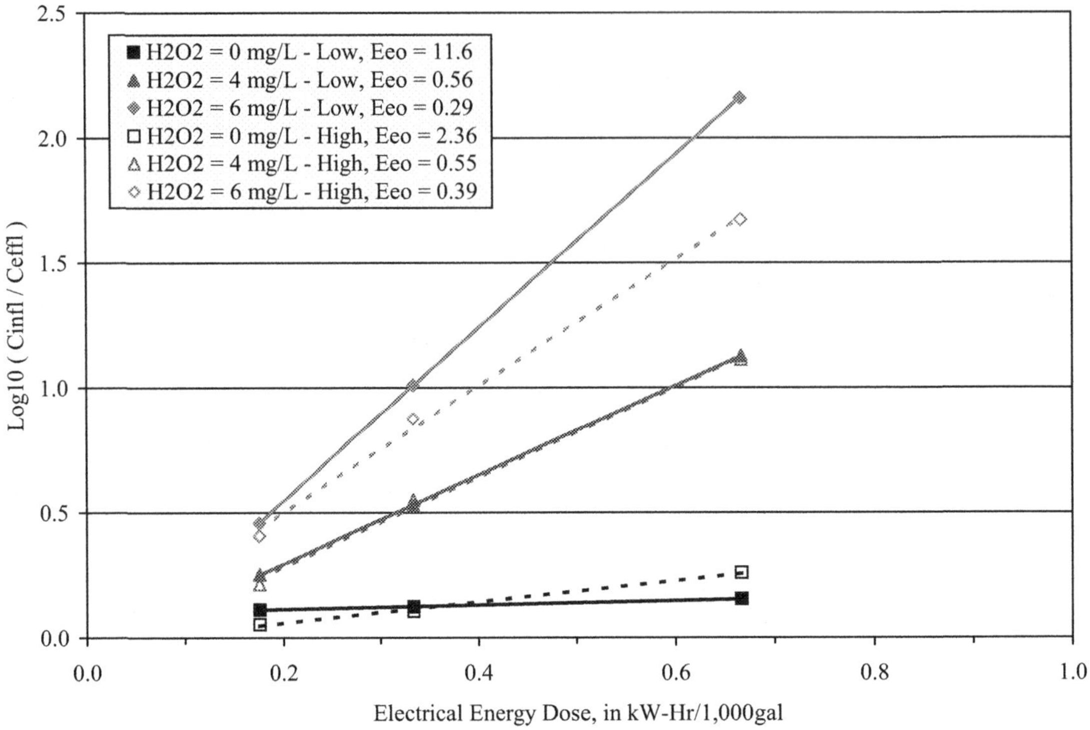

Figure 7.9 Plot of Log_{10} (C_{inf}/C_{eff}) of MIB vs. EED for the LPHO UV reactor

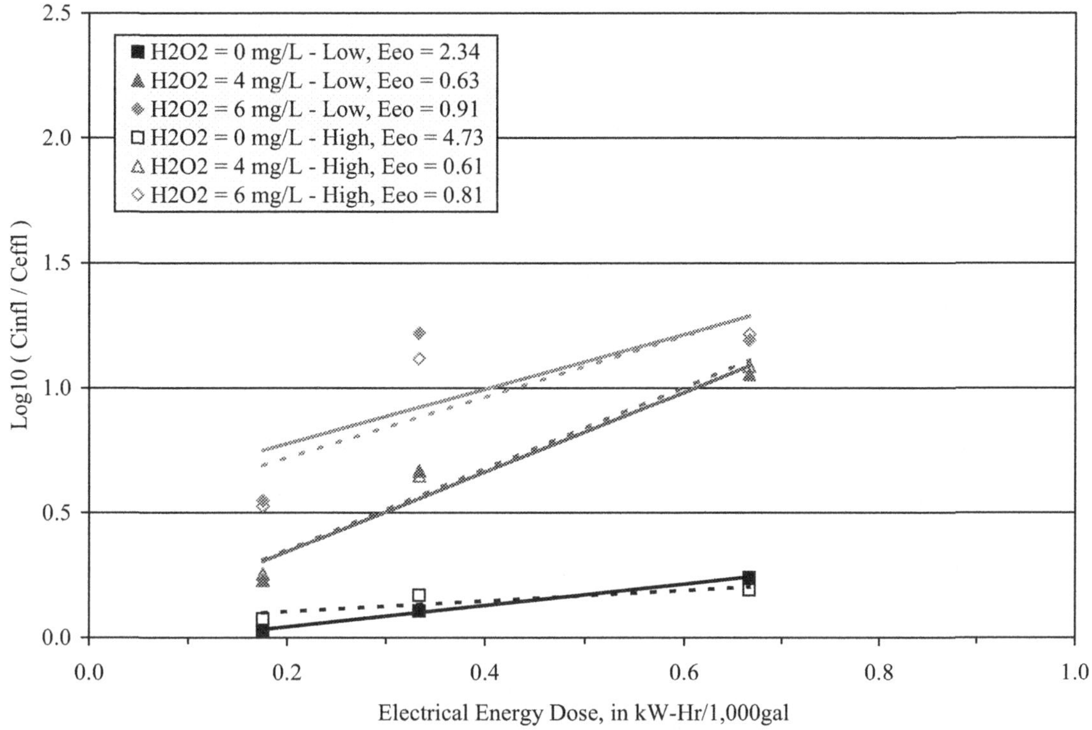

Figure 7.10 Plot of Log_{10} (C_{inf}/C_{eff}) of GSM vs. EED for the LPHO UV reactor

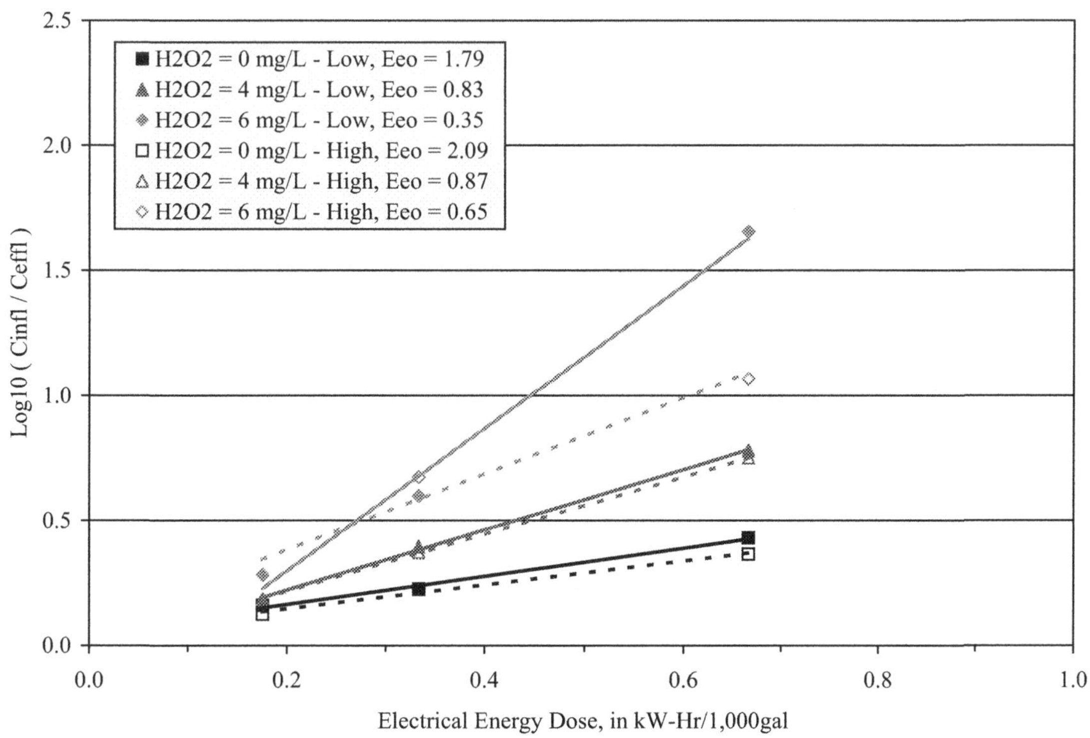

Figure 7.11 Plot of Log_{10} (C_{inf}/C_{eff}) of atrazine vs. EED for the LPHO UV reactor

Table 7.5
Summary of E_{EO} values for MIB, GSM and atrazine – LPHO UV reactor

H_2O_2, mg/L	Low or High Chemical Conc.	LPHO UV Reactor E_{EO} values, in kW-hr/1000gal/order		
		MIB	GSM	Atrazine
0	Low	11.6	2.34	1.79
4	Low	0.56	0.63	0.83
6	Low	0.29	0.91	0.35
0	High	2.36	4.73	2.09
4	High	0.55	0.61	0.87
6	High	0.39	0.81	0.65

NOTE: An order is defined as 1-log or 90 percent degradation.

A lower E_{EO} value implies that a lower electrical energy dose is required to achieve the required log removal. As Table 7.5 indicates, the E_{EO} values decreased with increasing H_2O_2, at both low and high level of chemical concentrations of MIB, GSM and atrazine. For MIB and GSM, the decrease in E_{EO} was dramatic with the addition of H_2O_2. For atrazine, the decrease in E_{EO} was gradual with increasing H_2O_2 dose.

MP UV Reactor

The percent degradation of MIB, GSM and atrazine in the MP UV reactor was plotted against the EED for the three H_2O_2 doses, and for the low and high chemical concentrations. Figures 7.12 through 7.14 are plots of percent degradation of MIB, GSM and atrazine versus EED, for the MP UV reactor.

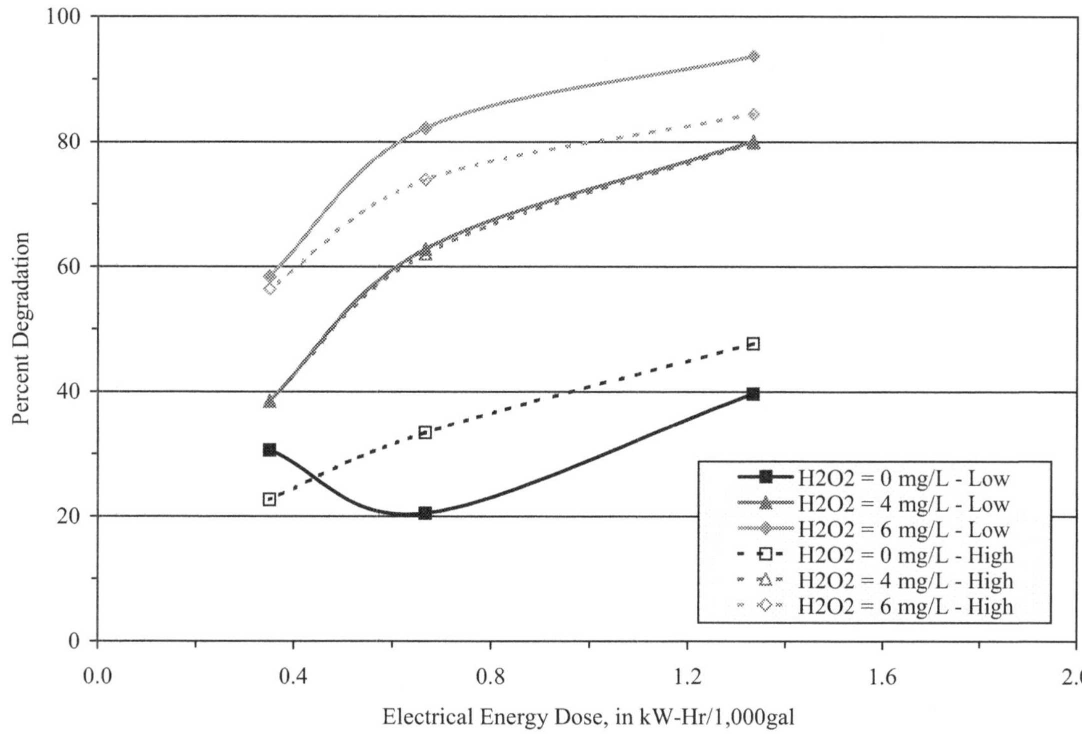

Figure 7.12 Percent degradation of MIB for the MP UV reactor

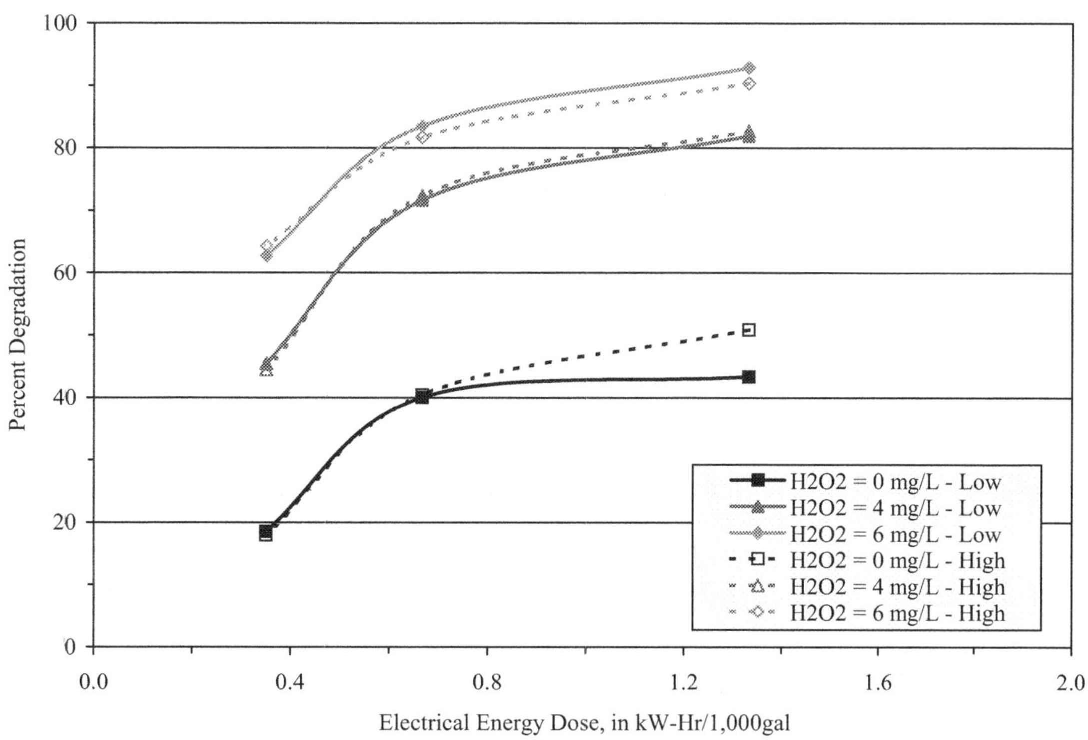

Figure 7.13 Percent degradation of GSM for the MP UV reactor

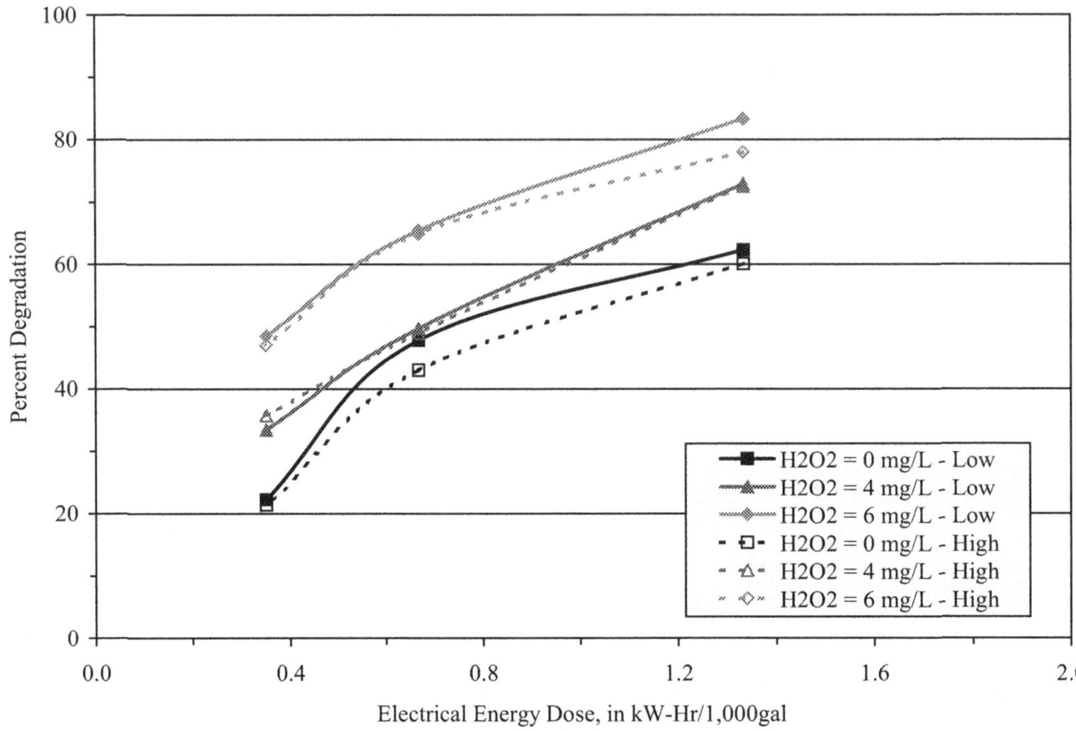

Figure 7.14 Percent degradation of atrazine for the MP UV reactor

For the MP UV reactor, the degradation of MIB, GSM and atrazine was directly proportional to the EED, and directly proportional to the H_2O_2 dose. In general, the concentration of the chemical did not affect the percent degradation. However, when H_2O_2 was not being added, there appeared to be some variation in percent degradation between the low and high levels.

For MIB and GSM, the percent degradation with no H_2O_2 being added was low (20 to 50 percent). With the addition of H_2O_2, there was a significant increase in MIB and geosmin degradation (rising to 40 to 95 percent).

Without the addition of H_2O_2, atrazine degradation was higher compared to MIB and GSM (20 to 60 percent). With the addition of H_2O_2, the increase in atrazine degradation was gradual (rising to 35 to 85 percent).

The E_{EO} values were calculated for the three chemicals (MIB, GSM and atrazine) at low and high concentrations, by plotting the logarithm (base 10) of influent concentration over effluent concentration against the electrical energy dose. A linear regression was fit to each data set. The E_{EO}'s were calculated from the inverse of the slope of these linear regression fits. Figures 7.15 through 7.17 are plots of Log_{10} (C_{inf}/C_{eff}) versus EED for the MP UV reactor, for MIB, GSM and atrazine. The E_{EO}'s are tabulated in Table 7.6.

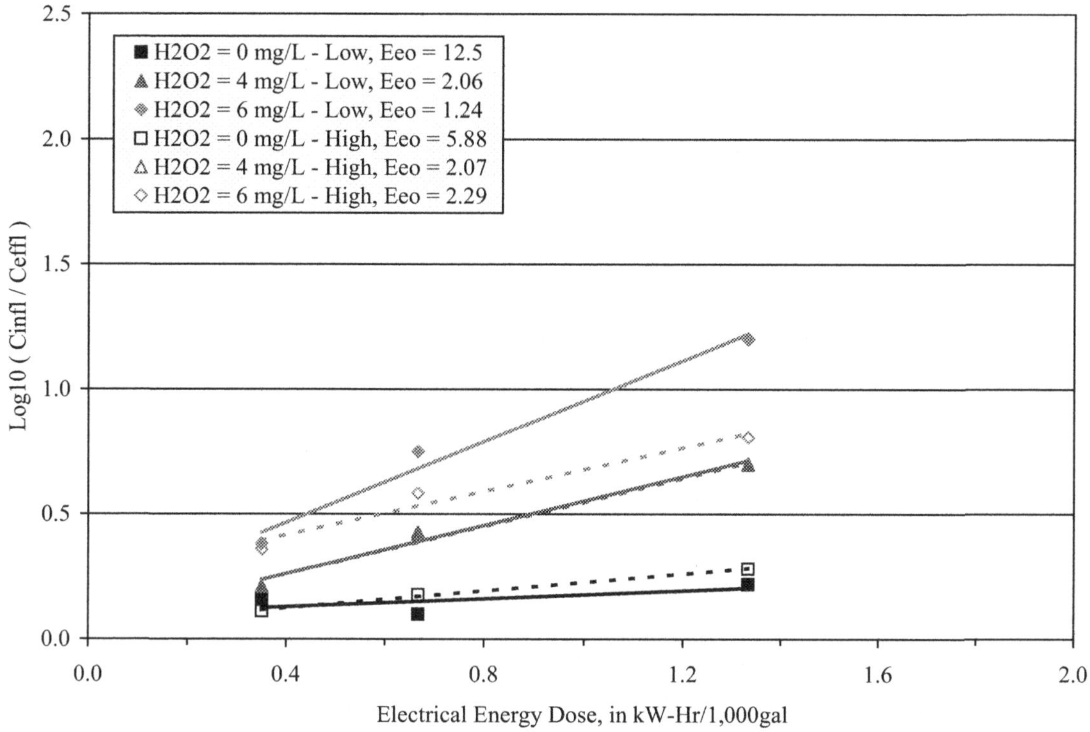

Figure 7.15 Plot of Log_{10} (C_{inf}/C_{eff}) of MIB vs. EED for the MP UV reactor

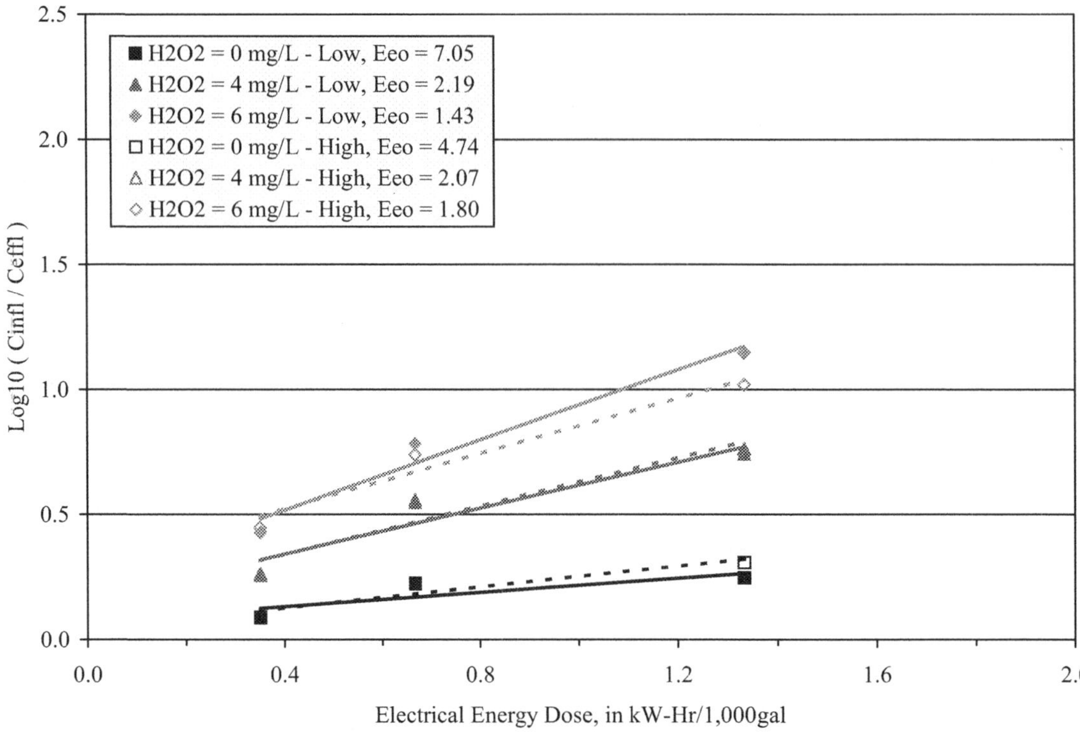

Figure 7.16 Plot of Log_{10} (C_{inf}/C_{eff}) of GSM vs. EED for the MP UV reactor

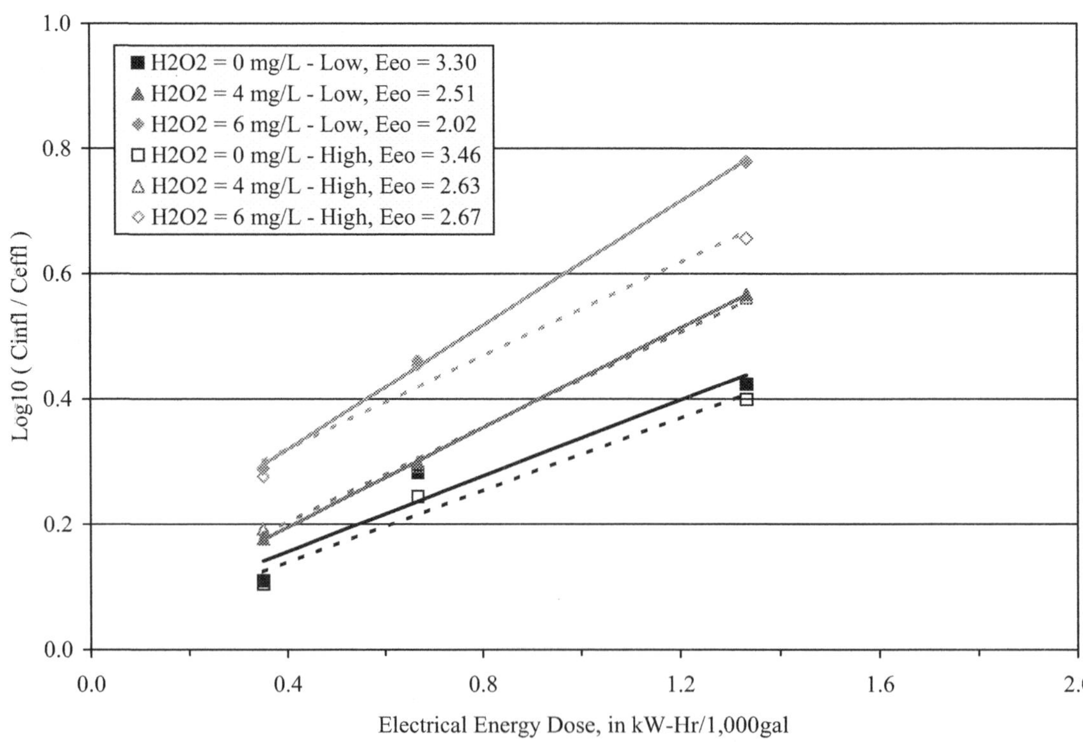

Figure 7.17 Plot of Log_{10} (C_{inf}/C_{eff}) of atrazine vs. EED for the MP UV reactor

Table 7.6
Summary of E_{EO} values for MIB, GSM and atrazine – MP UV reactor

H_2O_2, mg/L	Low or High Chemical Conc.	MP UV Reactor E_{EO} values, in kW-hr/1000gal/order		
		MIB	GSM	Atrazine
0	Low	12.5	7.05	3.30
4	Low	2.06	2.19	2.51
6	Low	1.24	1.43	2.02
0	High	5.88	4.74	3.46
4	High	2.07	2.07	2.63
6	High	2.29	1.80	2.67

For all three chemicals (MIB, GSM and atrazine), the E_{EO} values decreased with increasing H_2O_2 dose. At a H_2O_2 dose of 4 mg/L, the E_{EO} values at low and high levels of concentration were similar. However, at H_2O_2 doses of 0 and 6 mg/L, the E_{EO} values were different at low and high concentration levels. For MIB and GSM, the E_{EO} values dropped significantly with the addition of H_2O_2. For atrazine, the decrease in E_{EO} with the addition of H_2O_2 was gradual.

CHAPTER 8
CONCLUSIONS AND RECOMMENDATIONS

CONCLUSIONS

A number of conclusions may be drawn from this investigation of the UV photolysis and oxidation of the organic and organoleptic contaminants tested. The range of lamp types and contaminants investigated represents the current and emerging UV technologies available and the scope of chemical challenges that may utilities face. The research progressed focusing on a number of key questions from which to guide our research and obtain answers to – presented in the introduction chapter. As a means of conclusion, it is instructive go back to these questions and see what types of answers were reached as a result of the research efforts.

1. How do the doses necessary for contaminant degradation compare to those needed for UV disinfection?

Based on our research and the research of others before us, it is quite clear that at UV doses used for disinfection (approximately 40 mJ/cm^2) there will be almost no apparent chemical degradation benefits for pollutants of concern. In fact, for the contaminants researched in this study, to achieve one log of degradation via UV photolysis, the UV dose would have to be at least as high as 500 mJ/cm^2, as indicated in the bench-scale work. For some of the contaminants, doses even as high as 2,000 mJ/cm^2 were not effective for this level of degradation. Thus, the effectiveness for UV degradation is contaminant-specific. For some contaminants, such as NDMA, designing a UV system to achieve both disinfection and contaminant destruction would appear feasible. If an oxidant such as hydrogen peroxide is added to the water in the presence of UV light, then the UV reactivity of many compounds greatly increases.

2. How does water quality affect UV contaminant degradation treatment processes?

Based on the factorial statistical design utilized in this study, the water quality factors that most affected the UV based degradation processes tested were temperature and pH. However, it is important to note that the bench collimated beam tests were performed at UV doses that were "delivered" doses. Thus, the effect of such factors as TOC, which can alter the UV absorbance, was not counted in these effects but appeared to be of importance on the bench reactor scale and the pilot reactor scale. The addition of H_2O_2 played a more significant role for some contaminants (e.g. MIB) than for others (e.g., NDMA). On the pilot scale, the addition of H_2O_2 also played a major enhancing role for treatment of taste and odor compounds. Also the difference in the water quality between the waters tested at Detroit and Duke appeared to play some role in the different efficiencies of treatment for atrazine and taste and odor causing compounds, likely due to the differences in UV light screening and OH radical scavenging.

3. How does UV light from different sources affect treatment of different contaminants?

Three different UV sources were tested: LP, MP and P-UV lamps. Based on our analysis, on a per dose basis, the use of a MP UV source often improved treatment for most of the pollutants we tested. However, on the basis of electrical efficiency, the LP UV source was often the most efficient for degradation. We also tested two P-UV sources and, allowing for experimental uncertainties in the UV dose measurement for the pulsed lamps, they appeared to perform similar to or slightly better than the MP lamps for both atrazine and NDMA contaminants. Although the P-UV lamp results were intriguing, comparison with the mercury vapor lamps was complicated by several factors, including variable energy efficiency of different pulsed lamps and difficulties in accurately quantifying the UV dose for these lamps. We hope to overcome these obstacles in future research and pursue further testing of these non-mercury emerging light sources.

4. To what extent does UV0based advanced oxidation improve treatment for various contaminants compared to direct UV photolysis?

Addition of H_2O_2 leads to a marked increase in removal rates for atrazine with all the lamps, and adding oxidant decreased the electrical requirements for atrazine removal. With NDMA, H_2O_2 either had no effect or actually decreased the electrical efficiency of the process due to screening of the directly effective UV light. The treatment of MIB and geosmin was virtually non-existent without addition of H_2O_2. Thus, use of this additive is essential for UV treatment effectiveness against these contaminants. UV degradation of metolachlor, atrazine, diazinon, carbaryl, and toluene were all tested with and without H_2O_2 but these contaminants degraded much more rapidly in the presented of peroxide. In general, the addition of H_2O_2 in a UV pollutant treatment process greatly enhances the degradation of many contaminants.

5. Will UV form or affect the formation of disinfection byproducts?

This research provided an indepth investigation on the potential for UV both at doses used for disinfection and for pollutant degradation, to form direct disinfection byproducts or affect the formation of chlorinated byproducts following post chlorination. The formation of aldehydes and carboxylic acids as direct byproducts and trihalomethanes and haloacetic acids as post chlorination byproducts was tested. These products were monitored following UV treatment by LP, MP and P-UV sources in the presence and absence of H_2O_2 addition. UV treatment at doses that would be used for disinfection (up to 140 mJ/cm^2) caused no measurable changes in any of the byproducts for any of the lamp types. Furthermore, at doses used for pollutant degradation (~1,000-2,000 mJ/cm^2), there were no concerning changes in any of the disinfection byproducts measured. This was true in the presence or absence of H_2O_2.

At extremely high UV doses (>2,000 mJ/cm^2), there was a measurable decrease in the formation of THM's and HAA's following post-chlorination. Aldehydes were produced from each of LP UV, MP UV and P-UV irradiation, although significant increases in concentrations above background levels were not observed below applied doses of approximately 500 mJ/cm^2 for either UV technology. Above 500 mJ/cm^2, very low levels of aldehydes were formed (up to 13-45 µg/L at the maximum dose) under the polychromatic UV sources. In the presence of H_2O_2, the increases in aldehyde concentration began at UV doses greater than 140 mJ/cm^2. Carboxylic acids appeared to be produced from all LP UV, MP UV and P-UV irradiation, although significant increases in concentrations above background levels were not observed below doses of approximately 1,000 mJ/cm^2 for either MP LP UV. However, the increase in carboxylic acid concentrations with the P-UV doses varied more widely. In the presence of H_2O_2, the formation of carboxylic acids began at UV doses greater than 140 mJ/cm^2.

6. What are the relative efficiencies and costs for treating different contaminants with UV light?

Among the contaminants tested in the bench scale reactor, NDMA was by far the fastest to degrade under UV alone. Metolachlor was then the most reactive, followed by atrazine and diazinon, all of which had appreciable rates of direct photolysis. Carbaryl and toluene were both very unreactive with UV alone as were MIB and geosmin. With the addition of H_2O_2, all of these contaminants (except for NDMA in some cases) decayed much more rapidly. Interestingly, based on the relative E_{EO} values, the costs associated with treatment of the taste and odor compounds IPMP, geosmin, and MIB with UV/H_2O_2 will be very comparable to those for treating NDMA, both with LP and MP lamps. Of the remaining compounds, metolachlor is the most susceptible to UV/H_2O_2 treatment, and is more efficiently treated by LP/H_2O_2 than MP/H_2O_2. However, treating this compound will require approximately two to three times more electricity than NDMA for a given extent of removal. Much higher costs can be expected for treating ATZ and diazinon, and higher costs still for treating toluene and carbaryl.

RECOMMENDATIONS

Results presented throughout this report have provided insight into how to integrate UV based treatment of chemical contaminants into a water treatment plant as well as future needs for research. This section provides guidance and recommendations to scientists and utilities that may be interested in either better understanding UV degradation of contaminants or implementing UV technology for such applications. Recommendations include:

1. Designs for UV disinfection can include provisions for treatment of some photoreactive contaminants, but the UV doses necessary will be a factor of 5 to 10 times greater than for disinfection alone. The addition of H_2O_2 would decrease this factor for some contaminants.
2. UV treatment of contaminants will not be greatly affected by water quality factors such as pH and temperature. However, factors that alter the UV absorbance will have a significant effect on the applied dose and electrical power requirements for a UV treatment system. In a UV AOP process, the presence of OH radical scavengers and their effect on the treatment of the contaminants should be assessed through modeling or small scale studies.
3. The use of LP, MP or P-UV sources will depend upon the water quality, the type of pollutant being targeted and design factors such as layout and space constraints. However, based on the findings of this research, LP UV sources require lower electrical energy needs than MP UV for equivalent treatment. On a UV dose basis, the MP system typically outperformed the LP system, likely due to the greater absorbance of the polychromatic light by the pollutants investigated. The P-UV sources showed promise in the level of treatment achieved compared to the MP and LP sources, however, comparative overall costs for these systems was not investigated.
4. For many contaminants, the use of UV alone is just not feasible because of the high doses required for treatment. The potential for use of UV alone can be estimated with some fundamental information on the quantum yield of the pollutant in question. In many cases, it may be more effective to add H_2O_2 into the UV process as the treatment will be

much faster for some pollutants. The extent of treatment expected can be estimated using the OH radical rate constants for the contaminants and photochemical modeling. Based on the electrical efficiencies presented in the report, destruction of some contaminants using UV AOPs may be quite favorable. However, design and ultimate costs of these types of systems has to take into account costs of chemical additions (H_2O_2) and removal of the added chemicals.

5. In most cases, UV treatment can be implemented for disinfection and pollutant degradation without concern for production of disinfection byproducts. Infact, the use of UV AOPs at very high doses (>5000mJ/cm^2) can actually decrease the level of chlorinated byproducts formed.
6. Use of any oxidative treatment process must recognize that degradation of the parent contaminant does not necessarily guarantee removal of the biological or chemical activity of the pollutant by the breakdown products. Ultimate analysis must also include some measure of product water activity or degradation product identification to determine treatment effectiveness.

Practical Considerations From the Research

Many small and large utilities are already developing and applying UV technologies to their treatment processes as a control technology for *Cryptosporidium parvum* and other pathogens. However, there was little published research regarding the potential formation of UV DBPs and the oxidative capabilities of UV and AOPs for use in drinking water for the specific contaminants investigated. Given the results presented in this report, it appears safe to recognize that UV applied for disinfection will not either affect the formation of chlorinated disinfection byproducts during post-chlorination nor produce levels of aldehydes or carboxylic acids of public health concern.

Regarding the use of UV for contaminant degradation, this research has provided some fundamental bases for selection of different UV and UV/oxidation technologies given the different type of contaminants faced by the water industry. Clearly, there is no "one size fits all" answer with UV. For instance, comparing the treatment of NDMA, atrazine and MIB, an engineer may come up with similar or different UV based solutions for each contaminant type because each has a unique absorbance spectrum and quantum yield (relating to the potential for photolysis with different lamp types) as well as a unique hydroxyl radical reaction rate constant (relating to the effectiveness of UV/peroxide oxidation).

The side-by-side comparison of UV technologies on the bench and pilot scale for the ability of each to degrade compounds will aid designers in the types of analyses utilities will need to perform when deciding upon which technology to use. The comparative treatment efficiencies for different contaminants tested in the study help to recognize where some of the strengths and weaknesses of UV technology are with relation to certain pollutants. Thus, the knowledge generated can be used by both regulators and utilities alike in deciding how best to approach issues surrounding disinfection byproducts in drinking water and removal of unwanted organic and organoleptic compounds from water.

APPENDIX A
METHODS FOR QUENCHING H_2O_2 RESIDUALS PRIOR TO UFC TESTING

Hydrogen peroxide exerts a chlorine demand and, therefore, residual peroxide must be quenched prior to UFC testing and DBP analysis. A series of experiments was performed to determine the suitability of chlorine (sodium hypochlorite) and/or catalase for quenching hydrogen peroxide residuals in samples slated for UFC testing and DBP analysis. Residual peroxide had previously been quenched with chlorine with good success in the UW laboratory, however the extent of success depended on being able to accurately titrate the peroxide with a chlorine solution. A simpler method was in demand, and the use of catalase was suggested.

Initial concerns with catalase were that it would exert its own chlorine demand and, possibly, contribute to DBP formation. A series of catalase solutions with concentrations of between 0.05 and 4 mg/L were prepared, applied to 100 mg/L peroxide solutions and then analyzed by UFC testing and DBP analysis. Monitoring of the peroxide quenching kinetics of each test solution showed that even solutions as low in catalase concentration as 0.2 mg/L were effective in reducing the peroxide concentrations to below detection within 1 hour of application. Results of the UFC and DBP testing showed that catalase concentrations at or above 2 mg/L contributed measurably to chlorine demands and HAA yields.

Additional experiments were performed to compare the quenching effects of hydrogen peroxide by catalase and inorganic quenching agents including sodium hypochlorite, sodium sulphite and sodium thiosulphate.

The final result of this testing was that 0.2 mg/L catalase was chosen as the peroxide quenching agent for UFC experiments. It conveniently removed residual peroxide quickly, and did not contribute to the chlorine demand or DBP concentrations of samples into which it was added. Further details of the results have been reported in a paper by Liu et al. (2003).

APPENDIX B
SUMMARY OF STATISTICAL DATA FOR NDMA AND ATRAZINE WATER QUALITY EFFECTS

Table B.1
Summary of dose-based rate constants for ATZ degradation according to Stage 2 water quality effects matrix

STD Run	H_2O_2, ppm	pH	Temp	k (cm^2/mJ)	+/-
1	0.0	6.5	5.0	1.06E-03	7.25E-05
2	5.0	6.5	5.0	1.64E-03	2.01E-05
3	10	6.5	5.0	2.09E-03	7.92E-06
4	0.0	7.5	5.0	1.08E-03	9.95E-06
5	5.0	7.5	5.0	1.55E-03	1.34E-05
6	10	7.5	5.0	1.99E-03	4.97E-08
7	0	8.5	5.0	1.09E-03	2.37E-05
8	5.0	8.5	5.0	1.49E-03	3.88E-06
9	10	8.5	5.0	1.82E-03	8.92E-06
10	0	6.5	15.0	1.31E-03	2.96E-05
11	5.0	6.5	15.0	1.82E-03	1.36E-05
12	10	6.5	15.0	2.28E-03	5.29E-05
13	0.0	7.5	15.0	1.28E-03	9.44E-06
14	5.0	7.5	15.0	1.81E-03	3.03E-05
15	10	7.5	15.0	2.20E-03	3.02E-05
16	0	8.5	15.0	1.29E-03	1.17E-05
17	5.0	8.5	15.0	1.70E-03	1.78E-05
18	10	8.5	15.0	2.01E-03	5.14E-05
19	00	6.5	25.0	1.59E-03	2.51E-05
20	5.0	6.5	25.0	2.13E-03	4.05E-05
21	10	6.5	25.0	2.54E-03	2.38E-05
22	0	7.5	25.0	1.53E-03	1.18E-05
23	5.0	7.5	25.0	2.01E-03	1.63E-05
24	10	7.5	25.0	2.49E-03	8.79E-05
25	0.0	8.5	25.0	1.51E-03	3.53E-05
26	5.0	8.5	25.0	1.84E-03	3.87E-07
27	10	8.5	25.0	2.14E-03	4.03E-05
28	5.0	7.5	15.0	1.80E-03	-
29	5.0	7.5	15.0	1.73E-03	-
30	5.0	7.5	15.0	1.85E-03	-
31	5.0	7.5	15.0	1.81E-03	-
32	5.0	7.5	15.0	1.77E-03	-

The effect of peroxide, pH, and temperature on the rate constants of NDMA and ATZ was evaluated by multiple regressions, covariance coefficients, and by using correlation matrix. The rate constants correlated positively with peroxide and temperature and related inversely to pH.

Table B.2 provides summary statistical information for peroxide, pH, temperature, k_{NDMA}, and k_{ATZ}.

Table B.2
Summary Statistics for Parameters Evaluated

	Peroxide	*pH*	*Temp*	k_{NDMA}	k_{ATZ}
Mean	5.17	7.57	15.67	0.0041	0.0018
Standard Error	0.70	0.14	1.35	0.0001	0.0001
Median	5.00	7.50	15.00	0.0040	0.0018
Standard Deviation	3.82	0.74	7.40	0.0007	0.0004
Sample Variance	14.63	0.55	54.71	0.0000	0.0000
Kurtosis	-1.23	-1.09	-1.09	0.5098	-0.2755
Skewness	-0.06	-0.11	-0.11	0.7707	-0.0031
Range	10.00	2.00	20.00	0.0029	0.0015
Minimum	0.00	6.50	5.00	0.0031	0.0011
Maximum	10.00	8.50	25.00	0.0060	0.0025
Sum	155.00	227.00	470.00	0.1234	0.0535
Count	30.00	30.00	30.00	30.0000	30.0000
Largest(1)	10.00	8.50	25.00	0.0060	0.0025
Smallest(1)	0.00	6.50	5.00	0.0031	0.0011
Confidence Level(95.0%)	1.43	0.28	2.76	0.0002	0.0001

Tables B.3 and B.4 summarizes correlation and covariance constants for NDM and ATZ as a function of peroxide, pH, and temperature. Both Tables indicate a positive effect on peroxide and temperature and inverse relationship with pH.

Table B.3
Pearson Correlation Matrix

	H2O2	*pH*	*Temp*
NDMA	0.139875	-0.40085	0.830184
ATZ	0.865686	-0.17408	0.438667

Table B.4
Covariance Matrix

	H2O2	*pH*	*Temp*
NDMA	0.000265	-0.00026	0.003911
ATZ	0.001084	-7.8E-05	0.000942

Rate constants for NDMA and ATZ were regressed with experimental matrix data and the results are summarized in Tables B.5, B.6, and B.7.

Table B.5
Multiple Regression Constants

	NDA	ATZ
Multiple R	0.932442	0.985974
R Square	0.869449	0.972145
Adjusted R Square	0.855461	0.969161
Standard Error	0.000249	6.72E-05
Observations	32	32

Table B.6
ATZ – Multiple Linear Regression

	Coefficients	*Standard Error*	*t Stat*	*P-value*
Intercept	0.005477	0.000454	12.06479	1.31E-12
H2O2	2.4E-05	1.17E-05	2.048465	0.049994
pH	-0.00034	5.86E-05	-5.87037	2.6E-06
Temp	7.12E-05	5.86E-06	12.15802	1.09E-12

Table B.7
NDMA – Multiple Linear Regression

	Coefficients	*Standard Error*	*t Stat*	*P-value*
Intercept	0.00164776	0.000122723	13.42665	1.00671E-13
H2O2	8.6924E-05	3.16704E-06	27.44657	8.76043E-22
pH	-8.74E-05	1.58352E-05	-5.51925	6.72267E-06
Temp	2.2023E-05	1.58352E-06	13.90794	4.25253E-14

The coefficients obtained from this evaluation again indicate a positive effect of peroxide and temperature and an inverse effect on pH. This evaluation confirms conclusions drawn from correlation and covariance matrices.

APPENDIX C
EFFECTS OF HIGH DOSE MP UV IRRADIATION WITH OR WITHOUT H_2O_2 ON DBP FORMATION

In this section, DBP results from experiments employing high doses (above 1000 mJ/cm^2) of MP UV irradiation with or without H_2O_2 performed at UW are reported. These experiments were performed to further investigate the effects of UV irradiation under high UV doses on the formation of DBPs. Experiments in Phase 1 had indicated that high UV doses resulted in increases in aldehydes and carboxylic acids and decreases in THMs and HAAs.

The experiments were performed in UW laboratories using the MP UV collimated beam unit (Rayox lamp) simulating the MP UV experiments that had been performed at BPI and Duke. Samples of synthetic water (350 mL) were prepared according to the most recent recipe, placed in a 125x65 mm crystallization dish (sample depth approximately 3 cm) and irradiated with MP UV with or without H_2O_2. The UV doses were determined using an Excel spreadsheet (Fluence-MP-Rayox Lamp) provided by J.R. Bolton. Doses of approximately 0, 1000, 2000, 3000, 4000, and 5000 mJ/cm^2 were used in the experiments, and initial H_2O_2 concentrations were 98.8 mg/L, when applicable. When H_2O_2 was present in the sample, 0.2 mg/L catalase was added immediately after irradiation to quench the residual H_2O_2. Samples were refrigerated until DBPs were measured. The methods of UFC testing and DBP measurement were not changed.

The TTHM and aldehydes data are plotted in Figures C.1 and C.2 together with the data for similar experiments performed previously at Duke (designated as (A)) and BPI (designated as (B)).

In general, at high MP UV doses, the concentrations of aldehydes increased with increasing UV doses, whereas TTHMs and HAAs decreased with increases in UV doses. The presence of H_2O_2 augmented these trends. The following sections contain summaries of the results obtained.

Trihalomethanes

The effects of high doses (nominally 1000-5000 mJ/cm^2) of MP UV irradiation on TTHMs are shown in Figure C.1. The trends are similar to those of results reported in Chapter 5. For increases in MP UV dose from 968 mJ/cm^2 to 4841 mJ/cm^2 at increments of approximately 1000 mJ/cm^2, TTHM concentrations decreased linearly from 94.5 μg/L to 62.2 μg/L. When 98.8 mg/L H_2O_2 was added, TTHMs decreased linearly from 96.0 μg/L to 29.8 μg/L with UV doses increasing from 900 to 4501 mJ/cm^2. The results demonstrated that precursors of TTHMs were partly destroyed by UV irradiation at high UV doses, and the presence of H_2O_2 enhanced the destruction of TTHM precursors.

The concentrations of TTHMs in non-irradiated, chlorinated synthetic water were 115 and 102 μg/L, which are comparable to the results from the experiments performed previously at Duke and BPI. This confirmed the consistency of the experiments performed in different labs within the whole research group.

Aldehydes

Aldehydes produced from high MP UV doses increased linearly with increases in UV dose (Figure C.2). When MP UV doses increased from 968 mJ/cm^2 to 4841 at increments of

1000 mJ/cm^2, the aldehyde concentrations increased linearly from less than the detection limit to 56.9 μg/L. The presence of H_2O_2 (98.8 mg/L) did not change this trend appreciably.

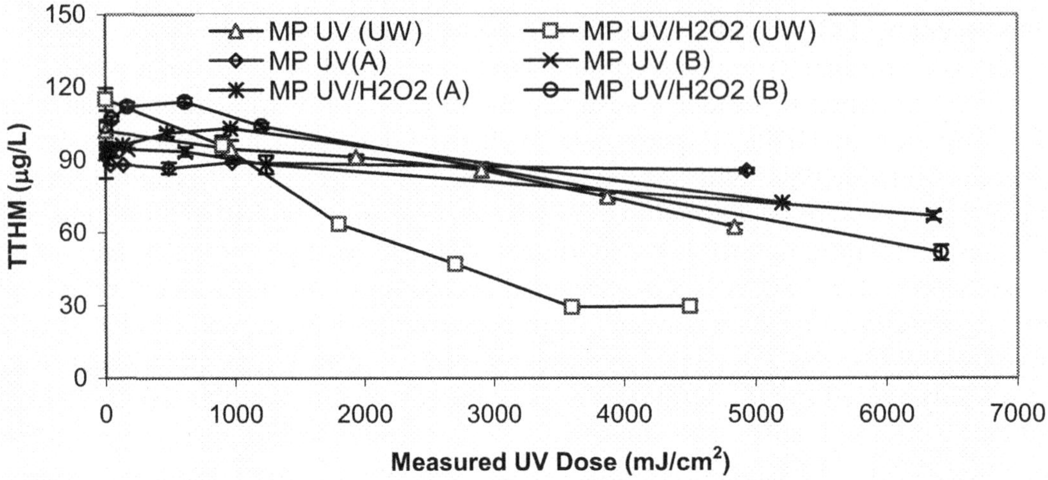

Figure C.1 TTHM concentrations from high MP UV dose on synthetic water in the presence or absence of 98.8 mg/L H_2O_2, followed by UFC testing (averages and standard deviations)

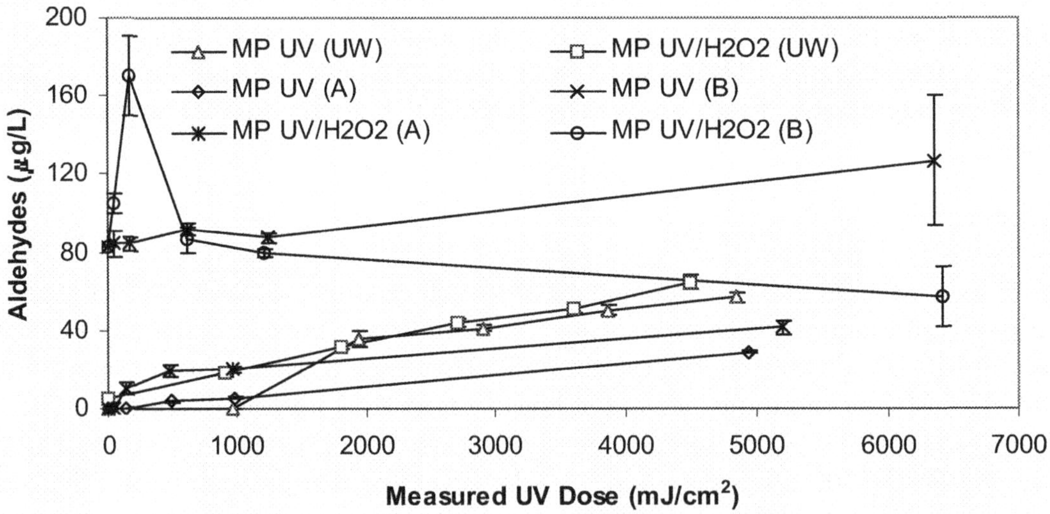

Figure C.2 Aldehyde concentrations from high MP UV dose on synthetic water in the presence or absence of 98.8 mg/L H_2O_2, followed by UFC testing (averages and standard deviations)

APPENDIX D
EFFECT OF TEMPERATURE ON DBP FORMATION

To address concerns of potential temperature effects on contaminant decomposition and DBP formation during UV irradiation and/or AOP applications, MP UV experiments were performed in Lab B at both a moderate temperature (25°C) and a low temperature (5°C). This section discusses the temperature effects on those DBPs that were generated directly from UV irradiation or AOPs (aldehydes and carboxylic acids) and the DBPs that were formed from UFC testing following UV irradiation or AOPs (HAAs and TTHMs).

Three factor-two level (2^3) factorial experiments were performed and the results are shown in Table D.1. The factorial analysis of the data contained in Table D.1 is shown in Table D.2. Generally, temperature variations caused only minor changes in the DBP yields in the UFC tests following MP UV irradiation/AOPs (TTHM and HAAs). There was a small, negative effect of temperature on TTHM and HAA generation, which is in agreement with the postulation that low temperature hindered the destruction of TTHM and HAA precursors thus allowing more TTHM and HAAs to be generated during subsequent UFC testing. In general, however, the results imply that the temperature of MP UV irradiation/AOPs (within the temperature range tested in this study) does not have an appreciable impact on TTHM and HAA precursors.

In contrast, the temperature conditions under which samples were irradiated by MP UV did have noticeable impacts on aldehyde and carboxylic acid concentrations which were generated directly from UV irradiation/AOPs. Contrary to the general knowledge that a higher temperature will facilitate chemical reaction, the experiment results in this section demonstrated that irradiation/AOPs at higher temperatures had a negative effect on aldehyde and carboxylic acid concentrations. The reasons for these results will be further investigated.

It was also found that among the three factors included in the factorial design, i.e. temperature, H_2O_2 dose and UV dose, UV dose was the most significant factor that impacted TTHM and aldehyde formation, while H_2O_2 impacted HAA and carboxylic acid concentrations the most. In addition, Table D.2 shows that excellent reproducibility was obtained for all of TTHM, HAA and aldehyde results, with standard deviations within ±5 μg/L, but generally less than ±3 μg/L. The carboxylic acid results were also reasonably reproducible.

Table D.1
Effect of temperature on DBP formation

Run	Temp. (°C)	H_2O_2 (mg/L)	UV Dose (mJ/cm^2)	TTHM (μg/L)	TTHM Std. Dev.	HAAs (μg/L)	HAAs Std. Dev.
1	5	0	0	91.7	0.0	89.9	3.4
2	5	100	0	93.8	4.0	106	2.0
3	25	0	0	93.1	0.0	108	1.6
4	25	100	0	97.0	1.5	112	0.4
5	5	0	1000	98.7	5.0	108	0.1
6	5	100	1000	111	2.8	129	1.2
7	25	0	1000	96.9	4.4	104	0.1
8	25	100	1000	95.9	1.0	107	0.6

Table D.2
Effect of temperature on DBP formation (Cont'd)

Run	Temp. (°C)	H_2O_2 (mg/L)	UV Dose (mJ/cm^2)	Aldehydes (μg/L)	Aldehydes Std. Dev.	Carboxylic Acids (μg/L)	Carboxylic Acids Std. Dev.
1	5	0	0	56.2	2.5	196	30.6
2	5	100	0	57.0	4.5	231	NA
3	25	0	0	52.0	0.7	205	22.9
4	25	100	0	60.5	3.0	185	18.2
5	5	0	1000	72.8	1.1	230	NA
6	5	100	1000	64.8	5.2	333	NA
7	25	0	1000	66.7	3.5	202	NA
8	25	100	1000	56.2	2.1	162	NA

Table D.3
Factorial analysis of temperature effects on DBP formation

Factor	Standardized Effect TTHM	HAAs	Aldehydes	Carboxylic Acids
Temperature	-0.149	-0.002	-0.244	-1.938
H_2O_2 dose	0.215	0.529	-0.133	0.520
UV dose	0.337	0.385	0.553	0.405

APPENDIX E
EFFECT OF BICARBONATE PRESENCE IN SYNTHETIC WATER ON THE FORMATION OF DBPS

This section deals with the potential effect of bicarbonate on the formation of DBPs by UV irradiation. The irradiation experiments were performed at Trojan laboratories using MP UV at doses of 500 and 5000 mJ/cm^2. Atrazine was added as a target contaminant and potential precursor of DBPs. H_2O_2 was also tested for its effect on DBP formation. The factors assessed in the experiments and the DBPs that were produced are shown in Table E.1. The methods for UV testing, water sample shipment, UFC testing and DBP measurement were the same as described in Chapter 5.

Effect of bicarbonate on the formation of DBPs at UV dose of 500 mJ/cm^2

The DBPs produced in the factorial test at a UV dose of 500 mJ/cm^2 are shown in Table E.1. The calculated main effects and interactions are summarized in Table E.2. Generally, bicarbonate had a significant impact on the formation of all DBPs other than HAA_9. The most significant impacts on DBP formation were associated with UV irradiation dose. Atrazine had the least effect on the four classes of DBPs, which allowed us to eliminate atrazine as a DBP precursor. It also showed that the effect of bicarbonate could not be ignored, and suggested that the water recipe should be carefully prepared when bicarbonate is needed. Also, in future testing using real source water, the water characteristics should be carefully analyzed in regard to its bicarbonate level.

Table E.1
DBPs produced from synthetic water with or without bicarbonate under UV irradiation at dose of 500 mJ/cm^2

Sample	Atrazine (μg/L)	H_2O_2 (mg/L)	UV (mJ/cm^2)	HCO_3^- (mg/L)	TTHM (μg/L)	HAAs (μg/L)	Alde. (μg/L)	C. Acids (μg/L)
1	0	0	0	0	92.5	82.6	66.7	LDL
2	102	0	0	0	93.1	80.7	52.6	LDL
3	0	102	0	0	93.1	83.5	43.6	LDL
4	100	102	0	0	92.9	82.7	39.8	LDL
5	0	0	493	0	95.9	77.9	47.6	67.0
6	102	0	493	0	96.1	76.6	43.3	64.2
7	0	102	536	0	108	80.4	89.7	266
8	100	102	536	0	108	71.9	68.3	123
9	0	0	0	126	87.6	71.9	33.2	99.4
10	103	0	0	126	88.7	77.8	27.8	66.2
11	0	107	0	126	93.9	80.4	30.5	72.1
12	98	107	0	126	88.0	80.7	28.4	72.4
13	0	0	515	126	123	82.7	35.6	83.6
14	103	0	515	126	119	83.5	31.1	101
15	0	107	555	126	145	76.3	42.1	148
16	98	107	555	126	144	82.6	41.1	147

LDL: Less than the Minimum Detection Limit

Table E.2
Main effect and interaction analysis of factorial testing (UV dose 500 mJ/cm^2)

Parameter	TTHM	HAA_9	Aldehydes	Carboxylic acids
Atrazine	-0.58	0.05	-3.54	-10.1
HCO^-_3	6.85	-0.03	-11.4	16.8
H_2O_2	4.81	0.30	2.85	21.7
UV	13.1	-0.53	4.76	43.1
Atrazine + HCO^-_3	-0.65	1.61	1.91	8.08
Atrazine + H_2O_2	-0.31	-0.39	0.00	-7.82
Atrazine + UV	-0.03	-0.39	-0.36	-6.03
HCO^-_3 + H_2O_2	1.76	0.21	-1.05	-10.5
HCO^-_3 + UV	8.53	2.31	-1.01	-21.9
H_2O_2 + UV	4.06	-1.49	7.60	24.3

The relative magnitudes of the main effects and 2-factor interactions of each of the four factors assessed in the experiments on the formation of the four classes of DBPs can be ranked as follows:

1) TTHMs. The UV dose was by far the most significant factor affecting TTHM concentrations (formed during subsequent UFC testing), with concentrations of bicarbonate and H_2O_2 producing less of an effect. The presence of atrazine at a concentration of 100 μM reduced the formation of TTHMs, although only slightly and so this effect might be ignored. The interactive effects of pairs of these factors on the formation TTHM were also determined and compared. The most significant interaction effect came from the combination of UV and bicarbonate, followed by the combination of UV and H_2O_2, and the combination of bicarbonate and H_2O_2. A model estimating these effects would take the following form:

2) HAA_9. Compared to their effect on the formation of TTHMs, UV dose, H_2O_2 concentration, Atrazine concentration, and bicarbonate concentration had little impact on the formation of HAA_9 (formed during subsequent UFC testing). This means that at the dose of 500 mJ/cm^2, UV did not change the structural characteristics of HAA precursors. Some of interactive effects of pairs of these factors were more significant than that of any single factor. The most significant interaction was produced by the combination of UV and bicarbonate, followed by the combination of Atrazine and bicarbonate, combination of UV and H_2O_2. Although the effect of single factors could be ignored, the interaction effect of some of these two factors could not be neglected.

3) Aldehydes. Each of the four factors had a definite impact on the formation of aldehydes. Of these four factors, bicarbonate had the most significant and negative effect on the formation of aldehydes, which means that increasing bicarbonate concentrations were associated with decreasing aldehyde concentrations. Increasing UV doses, however, increased aldehyde concentrations, and the positive impact of H_2O_2 on aldehyde formation was also nothing unexpected. It is interesting to note that Atrazine had a negative effect on aldehyde formation. The reason for this needs to be further investigated. The most significant interactive effect was the combination of H_2O_2 and UV, supporting that the UV/H_2O_2 advanced oxidation process enhanced the formation of aldehydes. Other interactions were relatively less significant.

4) Carboxylic acids: UV dose had the most significantly positive effect, followed by H_2O_2 and bicarbonate, whereas Atrazine had a negative effect on formation of carboxylic acids.

The most significantly positive interaction effect came from the combination of UV dose and H_2O_2, followed by the combination of bicarbonate and Atrazine. The combination of UV dose and bicarbonate had the most significant negative interaction, followed by the combinations of bicarbonate and H_2O_2, Atrazine and bicarbonate, and Atrazine and UV dose.

Effect of bicarbonate on the formation of DBPs at UV dose of 5000 mJ/cm^2

The DBPs generated in the factorial tests performed at the UV dose of 5000 mJ/cm^2 and the resulting main effects and interactions are shown in Table E.3 and Table E.4, respectively. Contrary to the discussion regarding to the results obtained at an MP UV dose of 500 mJ/cm^2, H_2O_2 was the factor that had the most significant impact on DBP formation, whereas bicarbonate and UV had similar but lesser impacts. Similarly, Atrazine was the factor that had the least effect. The experiment demonstrated that oxidation by H_2O_2 was enhanced at high UV doses. Similarly, however, the role of bicarbonate in impacting DBP formation could not be ignored.

Table E.3
DBPs produced from synthetic water samples with or without bicarbonate under UV irradiation at a dose of 5000 mJ/cm^2

Sample	Atrazine (μg/L)	H_2O_2 (mg/L)	UV (mJ/cm^2)	HCO_3^- (mg/L)	TTHM (μg/L)	HAAs (μg/L)	Alde. (μg/L)	C. Acids (μg/L)
1	0	0	0	0	118	115	57.3	113
2	100	0	0	0	121	112	77.1	129
3	0	100	0	0	131	115	50.9	152
4	100	100	0	0	116	115	45.8	91.2
5	0	0	5000	0	82.8	85.7	72.1	304
6	100	0	5000	0	83.1	88.9	90.3	289
7	0	100	5000	0	63.9	68.6	83.6	567
8	100	100	5000	0	61.8	65.8	58.9	554
9	0	0	0	126	118	131	64.3	149
10	100	0	0	126	120	93.8	74.3	70.9
11	0	100	0	126	124	87.3	58.1	82.6
12	100	100	0	126	118	87.6	56.4	86.2
13	0	0	5000	126	89.0	90.2	93.5	269
14	100	0	5000	126	83.8	84.5	85.5	291
15	0	100	5000	126	66.1	51.5	52.1	499
16	100	100	5000	126	62.3	83.4	41.9	455

Table E.4
Main effect and interaction analysis of factorial testing (UV dose 5,000 mJ/cm²)

Parameter	TTHM	HAA_9	Aldehydes	Carboxylic acids
Atrazine	-1.68	-0.83	-0.11	-10.6
HCO_3^-	0.23	-3.54	-0.62	-18.5
H_2O_2	-4.54	-7.93	-10.42	54.5
UV	-2.33	-1.49	0.59	14.7
Atrazine + HCO_3^-	0.05	-0.51	-1.13	-1.49
Atrazine + H_2O_2	-1.69	4.51	-5.11	-3.71
HCO_3^- + H_2O_2	-0.51	-3.28	-3.22	-11.7
H_2O_2 + UV	-0.60	-0.21	-0.27	6.08

The relative magnitudes of the main effects and 2-factor interactions of each of the four factors assessed in the experiments on the formation of the four classes of DBPs can be ranked as follows:

1) TTHMs: Contrary to the results at a UV dose of 500 mJ/cm², UV and H_2O_2 had significant negative effects on the formation of TTHMs (formed during subsequent UFC testing) for synthetic water irradiated at a UV dose of 5000 mJ/cm². These results are consistent with those of previous experiments in that high MP UV doses resulted in decreases in TTHMs. In the present experiments, H_2O_2 had a more significant effect than UV dose, which means at high H_2O_2 levels (approximately 100 mg/L), H_2O_2 enhances the destruction of TTHM precursors. Atrazine presence was observed to have a negative effect on the formation of TTHMs. The reason is not clear. The effect of bicarbonate on formation of TTHM could be ignored. The interactive effects of pairs of these factors can be ignored, except for that of the combination of Atrazine and H_2O_2.

2) HAA_9: Similar to the observations for TTHMs, above, H_2O_2 and UV dose had significant negative effects on the formation of HAA_9, which is consistent with results obtained in previous experiments. Similarly again, H_2O_2 greatly enhanced the destruction of HAA precursors. Bicarbonate also had a negative effect on formation of HAA_9 and the effect of Atrazine could be ignored. The interactive effect of the combination of Atrazine and H_2O_2 was significantly positive, whereas the effect of bicarbonate and H_2O_2 was significantly negative. Other interactions could be ignored.

3) Aldehydes. H_2O_2 had the greatest single effect, a negative effect, on formation of aldehydes in this set of experiments. Effects of other factors essentially could be ignored. The interactive effect of the combination of Atrazine and H_2O_2 was significantly negative, whereas the interactive effect of combination of Atrazine and bicarbonate was moderately negative. Other interactions could be ignored.

4) Carboxylic acids. H_2O_2 had the most significantly positive effect on the formation of carboxylic acids, followed by UV dose. These results are reasonable because carboxylic acids are byproducts of oxidation. Bicarbonate had a negative effect on the formation of carboxylic acids as did Atrazine, although to a lesser extent. The interactive effect of the combination of UV and H_2O_2 was significant and increased the concentrations of carboxylic acids above those that would be predicted considering single factor effects alone. Other 2-factor interactions reduced carboxylic acid concentrations, with relative rankings of: the combination of bicarbonate and H_2O_2, the combination of H_2O_2 and Atrazine, and the combination of Atrazine and bicarbonate.

APPENDIX F
EFFECT ANALYSIS OF FACTORIAL MATRIX OF DBP EXPERIMENTS (NDMA)

The experiments reported in this section investigated the effects of UV, TOC, alkalinity, H_2O_2, and temperature at two levels on the formation of DBPs. The effect of NDMA on generation of DBPs was also included. The irradiation experiments were performed at Trojan laboratories using MP UV. Again, the methods for UV testing, water sample shipment, UFC testing and DBP measurement were same as described in the previous Periodic Report.

Main effect and interaction analysis

Effect analysis for this factorial matrix of DBP experiments performed in the presence of NDMA is shown in Table F.1.

Table F.1
Effect analysis of factorial matrix of DBPs (in the presence of NDMA)

Parameter	TTHM	HAA_9	Aldehydes	Carboxylic acids
H_2O_2	8.91	-0.42	-0.04	31.6
pH	-12.1	-5.22	7.30	-72.6
Alkalinity	-0.83	-5.72	0.65	-24.8
TOC	106	124	0.83	32.6
Temperature	4.78	-9.18	-0.25	32.1
H_2O_2 + pH	3.93	-6.99	1.45	-60.0
H_2O_2 + TOC	5.21	5.17	3.43	11.8
pH + TOC	-12.1	-8.03	2.99	-39.0
pH + Temp	-4.89	-3.48	1.54	2.00
Alkalinity + Temp.	5.95	0.49	2.01	-45.9
TOC + Temperature	0.85	0.43	3.11	11.8

It was not unexpected that TOC played a significant role in impacting the formation of both TTHM and HAA_9 because TOC is representative of the precursors of THMs and HAA_9. The relative magnitudes of the main effects and 2-factor interactions of each of the five factors assessed in the experiments on the formation of the four classes of DBPs can be ranked as follows:

1) TTHMs. TOC had a significant impact on the formation of TTHMs during subsequent UFC testing: with increases in TOC levels, TTHMs also increased. H_2O_2 and temperature also positively impacted the formation of TTHMs and the effect of alkalinity could be ignored. At first glance, it may seem surprising that pH was observed to have a negative effect on the formation of TTHMs. This seems contrary to the common knowledge that increasing pH increases TTHM concentrations. However, it must be remembered that only the UV irradiation portion of the test was performed under different pH conditions – subsequent UFC testing was always performed at pH 8 ± 0.1. Therefore, the results simply indicate that there was greater UV-mediated destruction of THM precursors at high pH values. The most significant interactions were observed for the combination of pH and TOC, followed by alkalinity and temperature, H_2O_2 and TOC, and finally pH and temperature.

2) HAA_9. Similarly, TOC had the most significant effect on the formation of HAA_9. Similar to what was observed for THMs, pH also negatively impacted the formation of HAA_9 in

subsequent UFC testing. Both temperature and alkalinity had a negative effect on HAA formation, and the effect of H_2O_2 could be ignored. The most significant positive interactive effect came from the combination of TOC and H_2O_2. The most significant negative interaction came from the combination of pH and TOC, followed by the combinations of pH and H_2O_2, and pH and temperature.

3) Aldehydes. In this set of experiment, only pH had an appreciable effect on aldehyde formation. The effects of other single factors could all be ignored. It is interesting that all of the 2-factor interactions were positive and moderate on the formation of aldehydes.

4) Carboxylic acids. pH had the most significant and negative effect on the formation of carboxylic acids. TOC, temperature and H_2O_2 all had positive effects on carboxylic formation, whereas alkalinity had a negative effect. The most significant negative interactive effect came from the combination of H_2O_2 and pH, followed by the combinations of alkalinity and temperature, and pH and TOC. Alternatively, the combination of H_2O_2 and TOC and the combination of TOC and temperature had similar positive effects on carboxylic acid formation.

It is interesting that temperature had different impacts on the formation of THMs and HAA_9, positively for THMs and negatively for HAA_9. The reason should be further investigated.

Comparison of DBP formation with UV irradiation and in the dark (in presence of NDMA)

Three sets of experiments were performed to investigate the formation of DBPs with or without UV irradiation. The observations in Table F.2 show that UV irradiation at a dose of 1000 mJ/cm^2 did not have an appreciable impact on the subsequent formation of TTHM and HAA_9 from UFC testing. The data could also be interpreted that at or below this dose, UV irradiation did not affect the precursors of TTHM and HAA_9. However, UV irradiation did increase the concentration of aldehydes and carboxylic acids, with only one exception, and similar carboxylic acid concentrations were found under both UV irradiation and dark conditions.

Table F.2
Comparison of DBP formation with or without UV irradiation (in the presence of NDMA)

H_2O_2 (mg/L)	pH	Alkalinity (mg/L) as $CaCO_3$	TOC (mg/L)	Temp. (°C)	NDMA (μg/L)	UV Dose (mJ/cm^2)	TTHM (μg/L)	HAA_9 (μg/L)	Alde. (μg/L)	C. Acids (μg/L)
0	6.8	240	12	25	100	1000	289	303	17.0	268
0	6.8	240	12	25	100	0	248	297	6.7	137
7.5	7.9	140	7	15	100	1000	128	161	19.1	333
7.5	7.9	140	7	15	100	0	111	154	LDL	83.2
15	8.2	240	12	25	100	1000	260	284	47.3	115
15	8.2	240	12	25	100	0	262	237	LDL	103

LDL: Less than the Minimum Detection Limit

Comparison of DBP formation under UV irradiation with or without NDMA

Three sets of experiments (total of 9 runs) were conducted to investigate the impact of NDMA at concentration of 100 μg/L on DBP formation. Table F.3 shows that, with one exception, NDMA did not affect the formation of TTHM, HAA_9 and carboxylic acids considerably. However, NDMA resulted in some increases in aldehyde concentrations and so might be considered to be an aldehyde precursor.

Table F.3
Comparison of DBP formation under UV Irradiation with or without NDMA

H_2O_2 (mg/L)	pH	Alkalinity (mg/L) as $CaCO_3$	TOC (mg/L)	Temp. (°C)	NDMA (μg/L)	UV Dose (mJ/cm^2)	TTHM (μg/L)	HAA_9 (μg/L)	Alde. (μg/L)	C. Acids (μg/L)
0	6.8	240	12	25	100	1000	289	303	17.0	268
0	6.8	240	12	25	0	1000	250	285	LDL	283
15	8.2	40	2	25	100	1000	50.4	49.7	22.8	222
15	8.2	40	2	25	0	1000	44.0	56.6	LDL	131
7.5	7.9	140	7	15	100	1000	130	157	24.5	241
7.5	7.9	140	7	15	100	1000	126	168	35.9	294
7.5	7.9	140	7	15	100	1000	128	161	19.1	333
7.5	7.9	140	7	15	100	1000	131	160	30.8	357
7.5	7.9	140	7	15	0	1000	130	144	4.4	393

LDL: Less than the Minimum Detection Limit

APPENDIX G
EXAMPLE OF RAW DATA COLLECTED FOR DBP EVALUATIONS

The following set of DBP results were obtained from the MP-UV irradiation of synthetic water at Duke University on December 4 and 5, 2000. It has been included here to illustrate a typical data set obtained from the DBP analysis of UV-irradiated water samples, with and without subsequent Uniform Formation Condition (UFC) testing.

MP-UV Irradiation of Synthetic Water

Date/Place of UV irradiation: Dec. 04-05, 2000/Duke
Date of UFC Test: Dec. 08-09, 2000/UW
Date of measurement of DBPs: Dec. 07-13, 2000/UW

Sample	Measured UV Dose (mJ/cm^2)	Chlorine Demand (mg/L as Cl_2)		TTHM (μg/L)		HAA_9 (μg/L)		Aldehydes (μg/L)		Carboxylic Acids (μg/L)	
		Avg.	Std.Dev.	Avg.	Std.Dev.	Avg.	Std.Dev.	Avg.	Std.Dev.	Avg.	Std.Dev.
SWMP0	0	3.47	0.01	91.4	8.9	49.7	0.1	LDL	NA	5.1	6.7
SWMP40	38	3.45	0.01	88.0	0.2	51.2	0.1	LDL	NA	1.5	0.2
SWMP140	136	3.49	0.01	88.0	0.3	54.1	0.8	LDL	NA	1.8	1.1
SWMP500	486	3.54	0.01	86.6	2.3	49.1	0.3	3.6	0.4	4.5	0.8
SWMP1000	978	3.66	0.04	89.0	0.5	48.1	2.0	5.5	0.4	6.2	1.4
SWMP5000	4935	4.30	0.02	85.5	1.1	42.4	0.6	28.7	0.4	108	14.7

NA: Not Applicable, LDL: Less than MDL

MP-UV Irradiation of Synthetic Water

Date/Place of UV irradiation: Dec. 04-05, 2000/Duke
Date of UFC Test: Dec. 08-09, 2000/UW
Date of measurement of DBPs: Dec. 07-13, 2000/UW

Sample	Measured UV Dose (mJ/cm^2)	UFC Test Chlorine Concentrations: Initial (mg/L as Cl_2)	24 hr Residual (mg/L as Cl_2)	Demand (mg/L as Cl_2)	TTHM (μg/L)	HAA_9 (μg/L)	Aldehydes (μg/L)	Carboxylic Acids (μg/L)
SWMP01	0	4.50	1.04	3.46	97.7	49.8	LDL	0.4
SWMP02	0	4.50	1.03	3.47	85.1	49.6	LDL	9.8
Average	0	4.50	1.04	3.47	91.4	49.7	LDL	5.1
SWMP401	38	4.50	1.04	3.46	88.1	51.2	LDL	1.3
SWMP402	38	4.50	1.06	3.44	87.8	51.1	LDL	1.6
Average	38	4.50	1.05	3.45	88.0	51.2	LDL	1.5
SWMP1401	136	4.50	1.00	3.50	88.3	51.0	LDL	1.5
SWMP1402	136	4.50	1.02	3.48	87.7	50.9	LDL	0.8
SWMP1403	136	4.50	1.01	3.49	88.0	52.3	LDL	3.0
Average	136	4.50	1.01	3.49	88.0	54.1	LDL	1.8
SWMP5001	486	4.50	0.97	3.53	84.9	49.3	3.3	5.1
SWMP5002	486	4.50	0.95	3.55	88.2	48.9	3.9	3.9
Average	486	4.50	0.96	3.54	86.6	49.1	3.6	4.5
SWMP10001	978	4.50	0.80	3.70	88.5	47.2	5.0	6.3
SWMP10002	978	4.50	0.82	3.68	89.5	46.7	5.8	7.5
SWMP10003	978	4.50	0.90	3.60	89.0	50.4	5.6	4.8
Average	978	4.50	0.84	3.66	89.0	48.1	5.5	6.2
SWMP50001	4935	5.00	0.72	4.28	86.2	42.8	28.9	118
SWMP50002	4935	5.00	0.68	4.32	84.7	41.9	28.4	97.4
Average	4935	5.00	0.70	4.30	85.5	42.4	28.7	108

MDL=Minimun Detection Limit; LDL=Less than the MDL

THM Concentrations in Samples fromMP-UV Irradiation of Synthetic Water

Date/Place of UV irradiation: Dec. 04-05, 2000/Duke
Date of UFC Test: Dec. 08-09, 2000/UW
Date of measurement of THMs: Dec.11-12, 2000/UW

Sample	Measured UV Dose (mJ/cm^2)	UFC Test Chlorine Concentrations: Initial (mg/L as Cl_2)	24 hr (mg/L as Cl_2)	Demand (mg/L as Cl_2)	1,2-DBP Int. Std. Area x 10^4	$CHCl_3$ (μg/L)	$CHCl_2Br$ (μg/L)	$CHClBr_2$ (μg/L)	$CHBr_3$ (μg/L)	TTHM (μg/L)
SWMP01	0	4.50	1.04	3.46	2.58	97.7	LDL	LDL	LDL	97.7
SWMP02	0	4.50	1.03	3.47	2.18	85.1	LDL	LDL	LDL	85.1
Average	0	4.50	1.04	3.47	2.38	91.4	LDL	LDL	LDL	91.4
SWMP401	38	4.50	1.04	3.46	2.28	88.1	LDL	LDL	LDL	88.1
SWMP402	38	4.50	1.06	3.44	2.25	87.8	LDL	LDL	LDL	87.8
Average	38	4.50	1.05	3.45	2.27	88.0	LDL	LDL	LDL	88.0
SWMP1401	136	4.50	1.00	3.50	2.29	88.3	LDL	LDL	LDL	88.3
SWMP1402	136	4.50	1.02	3.48	2.24	87.7	LDL	LDL	LDL	87.7
SWMP1403	136	4.50	1.01	3.49	2.28	88.0	LDL	LDL	LDL	88.0
Average	136	4.50	1.01	3.49	2.27	88.0	LDL	LDL	LDL	88.0
SWMP5001	486	4.50	0.97	3.53	2.56	84.9	LDL	LDL	LDL	84.9
SWMP5002	486	4.50	0.95	3.55	2.32	88.2	LDL	LDL	LDL	88.2
Average	486	4.50	0.96	3.54	2.44	86.6	LDL	LDL	LDL	86.6
SWMP10001	978	4.50	0.80	3.70	2.34	88.5	LDL	LDL	LDL	88.5
SWMP10002	978	4.50	0.82	3.68	2.37	89.5	LDL	LDL	LDL	89.5
SWMP10003	978	4.50	0.90	3.60	2.30	89.0	LDL	LDL	LDL	89.0
Average	978	4.50	0.84	3.66	2.34	89.0	LDL	LDL	LDL	89.0
SWMP50001	4935	5.00	0.72	4.28	2.27	86.2	LDL	LDL	LDL	86.2
SWMP50002	4935	5.00	0.68	4.32	2.67	84.7	LDL	LDL	LDL	84.7
Average	4935	5.00	0.70	4.30	2.47	85.5	LDL	LDL	LDL	85.5
MDL		0.01	0.01	0.01		0.08	0.05	0.05	0.01	

MDL=Minimun Detection Limit; LDL=Less than the MDL

HAA9 Concentrations in Samples fromMP-UV Irradiationof Synthetic Water

Date/Place of UV irradiation: Dec. 04-05, 2000/Duke
Date of UFC Test: Dec. 08-09, 2000/UW
Date of measurement of HAA9: Dec.12-13 2000/UW

Sample	Measured UV Dose (mJ/cm^2)	UFC Test Chlorine Concentrations: Initial (mg/L as Cl_2)	UFC Test Chlorine Concentrations: 24 hr (mg/L as Cl_2)	UFC Test Chlorine Concentrations: Demand (mg/L as Cl_2)	1,2-DBP Int. Std. Area x 10^5	2,3,5,6TFBA Surrogate (530 μg/L)	DBPA Surrogate (51 μg/L)	ClAA (μg/L)	BrAA (μg/L)	Cl_2AA (μg/L)	Cl_3AA (μg/L)	BrClAA (μg/L)	$BrCl_2AA$ (μg/L)	Br_2AA (μg/L)	Br_2ClAA (μg/L)	Br_3AA (μg/L)	HAA_9 (μg/L)
SWMP01	0	4.50	1.04	3.46	1.84	504	38.7	LDL	LDL	26.8	18.0	5.0	LDL	LDL	LDL	LDL	49.8
SWMP02	0	4.50	1.03	3.47	1.75	505	39.3	LDL	LDL	26.2	18.2	5.2	LDL	LDL	LDL	LDL	49.6
Average	0	4.50	1.04	3.47	1.80	505	39.0	LDL	LDL	26.5	18.1	5.1	LDL	LDL	LDL	LDL	49.7
SWMP401	38	4.50	1.04	3.46	1.77	500	42.8	LDL	LDL	26.6	18.5	6.1	LDL	LDL	LDL	LDL	51.2
SWMP402	38	4.50	1.06	3.44	1.78	505	39.5	LDL	LDL	26.6	18.5	6.0	LDL	LDL	LDL	LDL	51.1
Average	38	4.50	1.05	3.45	1.78	503	41.2	LDL	LDL	26.6	18.5	6.1	LDL	LDL	LDL	LDL	51.2
SWMP1401	136	4.50	1.00	3.50	1.81	488	39.6	LDL	LDL	26.0	17.8	3.2	LDL	2.9	1.1	LDL	51.0
SWMP1402	136	4.50	1.02	3.48	1.83	492	42.9	LDL	LDL	26.6	18.6	5.7	LDL	LDL	LDL	LDL	50.9
SWMP1403	136	4.50	1.01	3.49	1.67	485	45.3	LDL	LDL	26.4	19.4	6.5	LDL	LDL	LDL	LDL	52.3
Average	136	4.50	1.01	3.49	1.77	488	42.6	LDL	LDL	26.3	18.6	5.1	LDL	2.9	1.1	LDL	54.1
SWMP5001	486	4.50	0.97	3.53	1.71	477	43.3	LDL	LDL	24.6	18.7	6.0	LDL	LDL	LDL	LDL	49.3
SWMP5002	486	4.50	0.95	3.55	1.69	477	43.6	LDL	LDL	25.0	18.2	5.7	LDL	LDL	LDL	LDL	48.9
Average	486	4.50	0.96	3.54	1.70	477	43.5	LDL	LDL	24.8	18.5	5.9	LDL	LDL	LDL	LDL	49.1
SWMP10001	978	4.50	0.80	3.70	1.67	470	43.2	LDL	LDL	24.4	17.1	5.7	LDL	LDL	LDL	LDL	47.2
SWMP10002	978	4.50	0.82	3.68	1.70	469	42.9	LDL	LDL	23.8	17.6	5.3	LDL	LDL	LDL	LDL	46.7
SWMP10003	978	4.50	0.90	3.60	1.68	475	44.0	LDL	LDL	25.9	18.4	6.1	LDL	LDL	LDL	LDL	50.4
Average	978	4.50	0.84	3.66	1.68	471	43.4	LDL	LDL	24.7	17.7	5.7	LDL	LDL	LDL	LDL	48.1
SWMP50001	4935	5.00	0.72	4.28	1.65	464	43.6	LDL	LDL	22.4	14.6	5.8	LDL	LDL	LDL	LDL	42.8
SWMP50002	4935	5.00	0.68	4.32	1.64	470	41.2	LDL	LDL	22.1	14.2	5.6	LDL	LDL	LDL	LDL	41.9
Average	4935	5.00	0.70	4.30	1.65	467	42.4	LDL	LDL	22.3	14.4	5.7	LDL	LDL	LDL	LDL	42.4
MDL		0.01	0.01	0.01				0.3	0.4	0.3	0.5	0.5	1.1	0.4	0.5	1.4	

MDL=Minimun Detection Limit; LDL=Less than the MDL

Aldehyde Concentrations in Samples fromMP-UV Irradiation of Synthetic Water

Date/Place of UV irradiation: Dec. 04-05, 2000/Duke
Date of measurement of Aldehydes: Dec. 07-08, 2000/UW

Sample	Measured UV Dose (mJ/cm^2)	1,2-DBP (area x 10^5)	Formaldehyde (µg/L)	Acetaldehyde (µg/L)	Glyoxal (µg/L)	Methyl Glyoxal (µg/L)	Total (µg/L)
SWMP01	0	1.65	LDL	LDL	LDL	LDL	LDL
SWMP02	0	1.64	LDL	LDL	LDL	LDL	LDL
Average	0	1.65	LDL	LDL	LDL	LDL	LDL
SWMP401	38	1.67	LDL	LDL	LDL	LDL	LDL
SWMP402	38	1.68	LDL	LDL	LDL	LDL	LDL
Average	38	1.68	LDL	LDL	LDL	LDL	LDL
SWMP1401	136	1.69	LDL	LDL	LDL	LDL	LDL
SWMP1402	136	1.70	LDL	LDL	LDL	LDL	LDL
SWMP1403	136	1.72	LDL	LDL	LDL	LDL	LDL
Average	136	1.70	LDL	LDL	LDL	LDL	LDL
SWMP5001	486	1.62	3.3	LDL	LDL	LDL	3.3
SWMP5002	486	1.64	3.9	LDL	LDL	LDL	3.9
Average	486	1.63	3.6	LDL	LDL	LDL	3.6
SWMP10001	978	1.65	5.0	LDL	LDL	LDL	5.0
SWMP10002	978	1.64	5.8	LDL	LDL	LDL	5.8
SWMP10003	978	1.63	5.6	LDL	LDL	LDL	5.6
Average	978	1.64	5.5	LDL	LDL	LDL	5.5
SWMP50001	4935	1.65	21.9	7.0	LDL	LDL	28.9
SWMP50002	4935	1.65	21.8	6.6	LDL	LDL	28.4
Average	4935	1.65	21.9	6.8	LDL	LDL	28.7
MDL			1.6	2.0	3.5	2.5	

MDL=Minimun Detection Limit; LDL=Less Than the MDL

Carboxylic Acid Concentrations in Samples fromMP-UV Irradiation of Synthetic Water

Date/Place of UV irradiation: Dec. 04-05, 2000/ Duke
Date of measurement of Carboxylic acids: Dec., 2000/UW

Sample	Measured UV Dose (mJ/cm^2)	Acetate (µg/L)	Formate (µg/L)	Oxalate (µg/L)	Total (µg/L)
SWMP01	0	LDL	0.4	LDL	0.4
SWMP02	0	LDL	9.8	LDL	9.8
Average	0	LDL	5.1	LDL	5.1
SWMP401	38	LDL	1.3	LDL	1.3
SWMP402	38	LDL	1.6	LDL	1.6
Average	38	LDL	1.5	LDL	1.5
SWMP1401	136	LDL	1.5	LDL	1.5
SWMP1402	136	LDL	0.8	LDL	0.8
SWMP1403	136	LDL	3.0	LDL	3.0
Average	136	LDL	1.8	LDL	1.8
SWMP5001	486	LDL	5.1	LDL	5.1
SWMP5002	486	LDL	3.9	LDL	3.9
Average	486	LDL	4.5	LDL	4.5
SWMP10001	978	LDL	6.3	LDL	6.3
SWMP10002	978	LDL	7.5	LDL	7.5
SWMP10003	978	LDL	4.8	LDL	4.8
Average	978	LDL	6.2	LDL	6.2
SWMP50001	4935	LDL	118	LDL	118
SWMP50002	4935	LDL	97.4	LDL	97.4
Average	4935	LDL	108	LDL	108
MDL		5.0	2.0	5.0	

MDL=Minimun Detection Limit; LDL=Less than the MDL

APPENDIX H
STATISTICAL ANALYSIS OF WATER TREATMENT FACTORS FOR THE REMOVAL OF 2-METHYLISOBORNEOL, GEOSMIN AND 2-ISOPROPYL-3-METHOXY PYRAZINE IN BENCH SCALE REACTOR AND PILOT REACTORS

INTRODUCTION

Ultraviolet (UV) radiation is being considered as a primary disinfectant for surface water, ground water and drinking water supplies in the United States by the U.S. Environmental Protection Agency. However, the effect of UV alone and advanced oxidation processes (AOPs), including ozone, hydrogen peroxide (H_2O_2) and UV and H_2O_2, to produce disinfection by-products and/or oxidize taste and odor compounds is unknown. This appendix is part of the larger project to determine the effectiveness of monochromatic and polychromatic UV technologies alone and with H_2O_2 to oxidize relevant organic compounds. This evaluation focuses on the effect of UV treatment on taste and odor compound removal in both a bench scale reactor and pilot plant reactor experiment and complements the statistical studies presented in Chapter 4.

Two of the most problematic taste and odor compounds in drinking water are 2-methylisoborneol (MIB) and geosmin. MIB imparts a musty odor while geosmin has an earthy odor. Both are metabolites of algae, such as cyanobacteria that bloom under specific environmental conditions in lakes and reservoirs. The third problematic taste and odor compound in drinking water is 2-isopropyl-3-methoxypyrazine (IPMP), which imparts a decaying vegetation and earthy odor. MIB, geosmin and IPMP have an odor threshold when dissolved in water of 9, 4 and 2 ng/L, respectively (AwwaRF 1987). These compounds are not harmful to human health but are aesthetically unpleasing and can be detected at low concentrations by humans. They are not removed by conventional water treatment processes and therefore activated carbon adsorption or hydroxyl radical oxidation is needed. Thus, UV alone is not sufficient to remove MIB, geosmin or IPMP, but a combination of UV with H_2O_2 is needed. Previous work on using AOPs for treatment of these compounds has been done by Glaze *et al.*, (1990) and Lalezary *et al.*, (1986).

OBJECTIVE

The objectives of this study were to

- Optimize various water treatment factors for the effectiveness of removing MIB, geosmin and IPMP at the bench scale and for MIB and geosmin at the pilot plant
- Determine the primary and secondary statistically significant combinations of the treatment factors
- Compare how well the pilot plant represents the bench scale experiment

HYPOTHESIS

It is hypothesized that a higher percentage of taste and odor compound removal will be seen when a combination of UV and H_2O_2 is used as an AOP. AOPs generate highly reactive intermediates such as the hydroxyl radical via UV radiation. These oxidizing intermediates can attack taste and odor compounds and help increase the degradation rate. The second hypothesis is a combination of a high H_2O_2 dose and low UV flow will further enhance compound degradation. The extra H_2O_2 and the additional time of UV and H_2O_2 interaction will generate more hydrogen radicals that can then attack the compounds. Also a longer exposure time will further enhance compound removal since there will be more time for the UV and H_2O_2 to interact. The last hypothesis is that of the three compounds, geosmin will be easiest to degrade due to less steric hindrance in the molecule which makes it easier for hydroxyl radicals to attack the compound.

METHODS

Water samples from bench scale reactor experiments at Duke University (See Figure 2.1) and from the pilot plant in Detroit (See Figures 7.2-7.4) were analyzed using solid-phase microextraction and gas chromatography-mass spectrometry (APHA 2000). The differences between the bench scale and pilot plant experiments are summarized in Table H.1.

Table H.1
Bench Scale and Pilot Plant Characteristics

	Treatment Variables	Compounds Analyzed
Bench Scale Batch Process	▪ Water (2) ▪ UV Lamp (2) ▪ H_20_2 (3) ▪ Time (3)	▪ MIB ▪ Geosmin ▪ IPMP
Pilot Plant Flow Process	▪ Reactor (2) ▪ UV Flow (3) ▪ H_20_2 Dose (3) ▪ Contaminant Concentration (2)	▪ MIB ▪ Geosmin

STATISTICAL ANALYSIS OF WATER TREATMENT VARIABLES BY AN ANALYSIS OF VARIANCE (ANOVA) USING THE S-PLUS PROGRAM

Bench Scale

Table H.2 summarizes the testing protocol and results for the bench scale experiment. Four treatment factors were looked at in various combinations to determine which combinations would be significant for compound removal. The factors tested were: water (Synthetic and Durham), UV lamp (Medium and Low pressure), H_2O_2 dose (0, 2 and 5 mg/L) and time (0, short and long). The run at 0 seconds was taken to be the control for that variable group. For instance, the run with a medium pressure UV lamp and synthetic water varied with either 2 or 5 mg/L of H_2O_2 and time of 0, 10 and 20 seconds. The run at 0 seconds was then taken as the control for that group.

Table H.2.
Bench Scale Protocol and Results

Sample	UV Lamp Pressure	Water Type	H2O2 (mg/L)	Time (sec)	MIB % removal	Geosmin % removal	IPMP % removal
MPS11	Medium	Synthetic	0	20	5	14	40
MPS12	Medium	Synthetic	0	20	74	77	54
MPH2S11	Medium	Synthetic	2	10	66	61	41
MPH2S12	Medium	Synthetic	2	10	0	21	65
MPH2S21	Medium	Synthetic	2	20	62	68	61
MPH2S22	Medium	Synthetic	2	20	47	74	55
MPH5S11	Medium	Synthetic	5	10	73	75	58
*MPH5S21	Medium	Synthetic	5	20	84	90	86
MPH2D11	Medium	Durham	2	7	42	37	30
MPH2D12	Medium	Durham	2	7	68	82	64
MPH2D21	Medium	Durham	2	14	82	83	60
*MPH2D22	Medium	Durham	2	14	88	95	85
MPH5D12	Medium	Durham	5	7	84	93	72
MPH5D22	Medium	Durham	5	14	88	100	96
LPS12	Low	Synthetic	0	240	0	8	0
LPS22	Low	Synthetic	0	480	11	3	0
LPH2S12	Low	Synthetic	2	240	42	55	18
LPH2S22	Low	Synthetic	2	480	60	79	49
LPH5S12	Low	Synthetic	5	240	74	83	45
*LPH5S22	Low	Synthetic	5	480	80	92	76
LPH2D11	Low	Durham	2	210	0	0	5
LPH2D21	Low	Durham	2	420	34	61	65
LPH5D11	Low	Durham	5	210	52	73	48
***LPH5D21**	Low	Durham	5	420	53	97	84

Note 1: MP and LP: Medium Pressure and Low Pressure UV lamp; H2 and H5: 2 mg/L and 5 mg/L of H_2O_2
S and D: Synthetic and Durham Water; #: refers to time and replicate; Note 2: Samples with * are best removed

Pilot Plant

Table H.3 and H.4 summarize the testing protocol and results for the pilot plant experiment. There were also four variables for the pilot plant: reactor (Calgon and Trojan), UV flow (50, 100 and 190 gpm), H_2O_2 dose (0, 4 and 6 mg/L) and contaminant concentration (150 and 300 ng/L). The Calgon reactor corresponds to a medium pressure UV while the Trojan reactor is a low pressure UV. Each run contained one influent and two effluent samples. The influent sample was taken before treatment while the Calgon and Trojan effluent samples were taken after going through the individual reactors with treatment.

Table H.3
MIB Pilot Plant Protocol and Results

Trojan Reactor Flow (gpm)	Calgon Reactor Flow (gpm)	H2O2 (mg/L)	MIB (ng/L)	Influent (ng/L)	Avg Observed MIB Conc.	Trojan Effluent	Percent Removal Trojan	Calgon Effluent	Percent Removal Calgon
50	50	0	0	--		--	--	--	--
50	50	0	150	156	150	94	37	79	48
50	50	0	150	156		121	20	92	39
100	100	0	150	146		129	14	91	39
190	--	0	150	171		143	5	--	--
--	190	0	150	128		--	--	100	33
50	50	0	300	238	313	213	32	137	56
100	100	0	300	329		282	10	191	39
190	--	0	300	345		251	20	--	--
--	190	0	300	338		--	--	290	7
50	50	4	150	184	153	21	86	51	67
50	50	4	150	184		16	90	38	75
100	100	4	150	125		40	74	68	56
190	--	4	150	158		75	51	--	--
--	190	4	150	146		--	--	126	18
50	50	4	300	295	306	33	89	73	76
100	100	4	300	360		87	71	114	63
100	100	4	300	360		117	62	--	--
190	--	4	300	298		163	47	--	--
--	190	4	300	305		--	--	178	42
50	50	6	150	121	147	<DL	>93	17	88
*100	100	6	150	173		11	92	15	90
190	--	6	150	152		57	61	--	--
--	190	6	150	142		--	--	46	69
50	50	6	300	333	371	14	96	62	83
50	50	6	300	333		--	--	35	91
50	50	6	300	333		--	--	49	87
100	100	6	300	349		41	89	69	81
190	--	6	300	468		131	65	--	--
--	190	6	300	332		--	--	118	68
50	50	4	300	269	307	30	90	55	82
50	50	4	300	342		19	94	56	82
50	50	4	300	309		25	92	80	74

Note 1: Detection limit (DL) of MIB is 10 ng/L

Note 2: Calculation of % removal at DL is 33-DL = (33-10)/33 = 23/33 = >69% removed

Note 3: Sample with * is best removed

Table H.4
Geosmin Pilot Plant Protocol and Results

Trojan Reactor Flow (gpm)	Calgon Reactor Flow (gpm)	H2O2 (mg/L)	Geos-min (ng/L)	Influent	Avg Observed Geosmin Conc.	Trojan Effl.	Percent Remov-al Trojan	Calgon Effl.	Percent Removal Calgon
50	50	0.0	0	--		--	--	--	--
*50	50	0.0	150	109	95	68	28	<DL	>89
100	100	0.0	150	145		<DL	>89	116	0
190	0	0.0	150	62		75	22	--	--
0	200	0.0	150	62		--	--	75	22
50	50	0.0	300	256	252			130	49
100	100	0.0	300	301		302	0	270	0
190	0	0.0	300	225		191	24	--	--
0	200	0.0	300	225		--	--	185	27
*50	50	4.9	150	168	123	<DL	>92	30	75
*100	100	4.9	150	100		<DL	>92	38	69
190	0	4.9	150	108		45	63	--	--
0	190	4.9	150	119		--	--	<DL	>92
50	50	4.9	300	78	134	17	87	33	76
100	100	4.9	300	106		20	85	42	68
190	0	4.9	300	102		44	67	--	--
0	195	4.9	300	249		--	--	40	70
*50	50	7.4	150	26	62	<DL	>84	<DL	>84
*100	100	7.4	150	32		<DL	>84	<DL	>84
*190	0	7.4	150	66		<DL	>84	--	--
0	195	7.4	150	122		--	--	29	53
50	50	7.4	300	246	264	47	82	48	82
100	100	7.4	300	256		78	71	89	66
190	0	7.4	300	306		90	66	--	--
0	195	7.4	300	249		--	--	46	82

Note 1: Detection limit (DL) of Geosmin is 10 ng/L
Note 2: Calculation of % removal at DL is 33-DL = (33-10)/33 = 23/33 = >69% removed
Note 3: Samples with * are best removed

ANALYSIS

Bench Scale

The bench scale experiment consisted of samples run at four experimental variables with controls at a time of 0 seconds. The controls denoted samples without treatment. The percent removal of MIB, geosmin and IPMP was then calculated by:

[(control run-experimental run) / (control run)] * 100

Each sample was run in duplicate and therefore in the percent removal calculation, the samples that ended in 1 were grouped together and the samples that ended in 2 were grouped together.

An Analysis of Variance (ANOVA) was then performed on the data set to see which variables significantly removed MIB, Geosmin and IPMP. The ANOVA model was used under the assumptions that the duplicates were considered to be independent and the residual errors were normally distributed with a mean of zero and constant variance. The normality of the residual errors was assessed in two ways. First, diagnostic plots were visually inspected. Then the one sample Kolmogorov-Smirnov Test of Composite Normality was run. The null hypothesis is that the data are normally distributed and if the p-value < 0.05 the null hypothesis is rejected. If the p-value > 0.05 then sufficient evidence is not available to reject the null hypothesis and it can be said that the residuals are normally distributed. Both of these methods were performed on each data set in order to see if the ANOVA model could be used.

A linear model was then constructed from the significant ANOVA variables which assisted in the prediction of average percent removal. This is an additive model where percent removal is the dependent variable and water, UV Lamp, H_2O_2 dose and time are independent variables, unless interactions are present where they will be in addition to their own main effects. The basic formula for the main effects of the bench scale linear model is:

$$\% \text{ removal} = a + b_1[\text{UV Lamp}] + b_2[\text{Water}] + b_3[H_2O_2] + b_4[\text{time}]$$

Where a is the baseline/intercept and is the average percent removal when all significant factors are at their low conditions of Durham water, low UV lamp, 0 mg/L of H_2O_2 and time 0. IPMP was an exception where Synthetic water was included in the baseline instead of Durham water. This was because all the treatment factors were significant in the initial ANOVA.

MIB

The MIB ANOVA gave UV Lamp and H_2O_2 dose as significant treatment effects, see Table H.5. The significance was based on the F-statistic having a p-value less than the alpha value of 0.05. This indicates that the treatment variables significantly reduced MIB percent removal. Before we could proceed to the linear regression model, the Kolmogorov-Smirnov test of Composite Normality was done along with the visual inspections of the diagnostic plots. The p-value for the test was 0.5 and therefore there was not enough evidence to reject the null hypothesis of the residuals being normally distributed. The assumption of residual normality was confirmed and the linear model was then generated from the significant ANOVA variables, see Table H.6.

Table H.5
Bench Scale MIB ANOVA

Terms added sequentially (first to last)	Df	Sum of Sq	Mean Sq	F Value	Pr(F)
UV.Lamp	**1**	**2554.5**	**2554.5**	**4.70**	**0.044**
Water	1	605.1	605.1	1.11	0.305
H2O2	**2**	**6977.2**	**3488.6**	**6.42**	**0.008**
Time	1	840.4	840.4	1.55	0.229
Residuals	18	9773.6	543.0		

Table H.6
Linear Regression Model for MIB Removal

	Value	Std. Error	t value	Pr(>|t|)
Intercept	11.1	12.5	0.89	0.385
UV.Lamp	**22.8**	**9.7**	**2.35**	**0.029**
$H_2O_2$2	23.0	13.4	1.72	0.101
$H_2O_2$5	**50.8**	**14.1**	**3.60**	**0.002**

The linear regression model shows that on average there is an 11.1% removal of MIB at the low significant values of low UV lamp and 0 mg/L of H_2O_2. However, when the UV is changed to medium pressure there is an additional 22.8% removal which increases the total average removal to 33.9%. In addition, if the H_2O_2 is changed to 5 mg/L an extra 50.8% removal is seen. This now increases the total average percent removal to 84.7%; however this additive linear model explains only 48.9% of the data variability. These results can be seen graphically in Figure H.1.

The graph clearly shows that the highest average percent removal is achieved with a medium pressure UV and 5 mg/L of H_2O_2. In addition, there seems to be a difference between 0 and 5 mg/L of H_2O_2 but it is not statistically significant since the error bars (representing $\pm$ 2 Standard Errors of the mean) on the experimental conditions for these two H_2O_2 doses overlap. The p-values in the linear regression model give further confirmation that a medium UV Lamp and H_2O_2 dose of 5 mg/L are significant in MIB degradation at the bench scale.

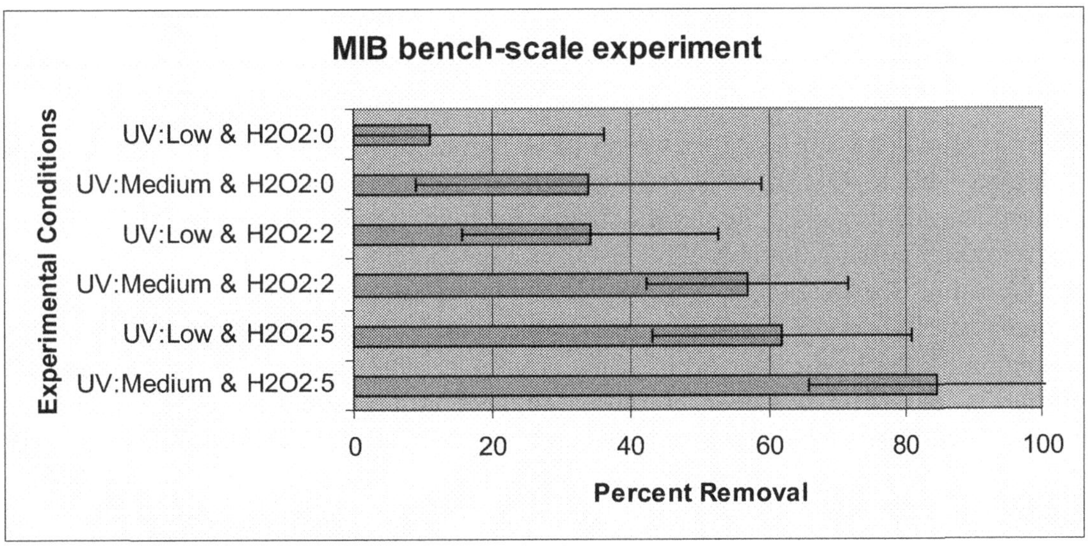

Figure H.1 MIB Percent Removal according to the Linear Regression Model

Geosmin

The same analytical procedure was completed for Geosmin at the bench scale and the ANOVA can be seen in Table H.7. Just the main effects were included since the visual inspections of the interaction plots did not show the presence of significant interactions. Only H_2O_2 had a p-value less than the alpha value of 0.05 in the ANOVA. Even though H_2O_2 dose was the only significant variable in the ANOVA, UV lamp was included into the linear model formula because it was a variable of interest. The diagnostic plots were inspected for normality and the Kolmogorov-Smirnov test of Composite Normality was done to verify the ANOVA's assumption. The p-value obtained was 0.5, which again does not provide sufficient evidence to reject the null hypothesis of normality for the residuals. The linear model was then generated, as presented in Table H.8.

Table H.7
Bench Scale Geosmin ANOVA

Terms added sequentially (first to last)					
	Df	Sum of Sq	Mean Sq	F Value	Pr(F)
UV.Lamp	1	1180.1	1180.1	2.15	0.160
Water	1	1250.4	1250.4	2.28	0.149
H2O2	**2**	**9841.0**	**4920.5**	**8.96**	**0.002**
Time	1	1515.0	1515.0	2.76	0.114
Residuals	18	9886.3	549.2		

Table H.8
Linear Regression Model for Geosmin Removal

	Value	Std. Error	t value	Pr(>|t|)
Intercept	17.5	13.0	1.35	0.193
UV.Lamp	16.0	10.0	1.59	0.128
H2O22	**31.6**	**13.9**	**2.28**	**0.034**
H2O25	**62.6**	**14.6**	**4.28**	**0.000**

On average there is a 17.5% removal of Geosmin at the low level of 0 mg/L of H_2O_2 and low UV lamp. The ANOVA had already shown, UV lamp is not a significant treatment factor for Geosmin degradation since the p-value is greater than the alpha-value of 0.05. However, the linear regression model illustrates that H_2O_2 is significant on all levels and it considerably enhances Geosmin degradation at 5 mg/L. There is an additional 62.6% removal of Geosmin when 5 mg/L of H_2O_2 is used which increases the total average removal to 80.1%. This linear model explains 51.7% of the variability. The graphical representation of the Geosmin model is very similar to the MIB bench-scale model representation, see Figure H.2. However, the Geosmin graph confirms the difference between 0 and 5 mg/L of H_2O_2 since the error bars do not overlap.

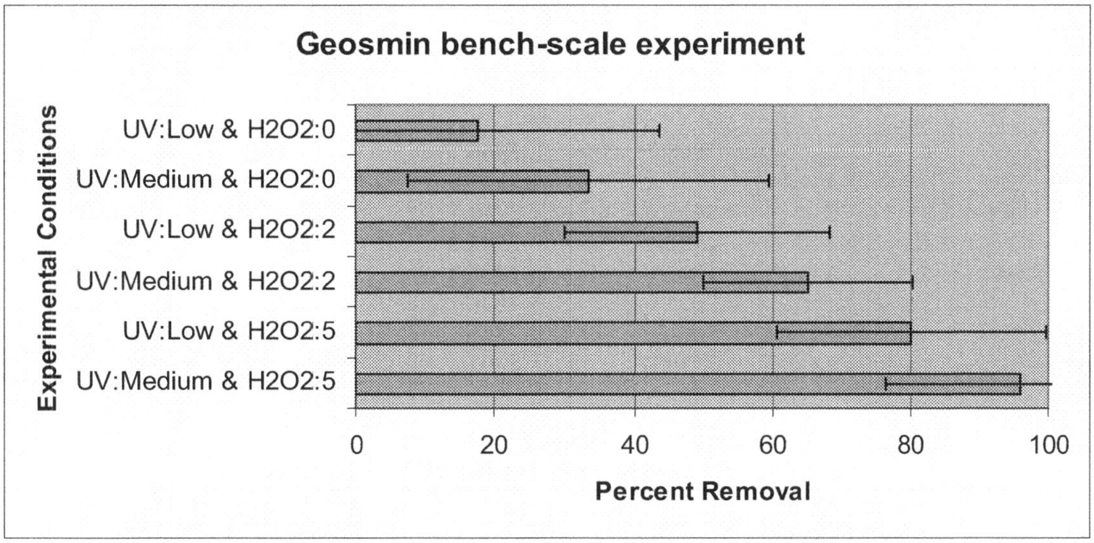

Figure H.2 Geosmin Percent Removal according to the Linear Regression Model

IPMP

IPMP percent removal was more complicated to model. This was because the main effects ANOVA showed all treatment variables as significant since the p-values were less than the alpha-value of 0.05, see Table H.9. The ANOVA was recalculated to include interactions which can be seen in Table H.10. Once more all the treatment variables were significant but only the UV lamp:time interaction was assessed as significant. The interaction is significant because the low pressure UV lamp was paired with a time over 200 seconds and the medium pressure UV lamp was paired with a time less then 30 seconds. This made the linear regression model more complex since an interaction would be taken into consideration along with the main effects. Visual inspections of the diagnostic plots for IPMP were done and the test of Composite Normality was also completed. The Kolmogorov-Smirnov test gave a p-value of 0.5 and therefore sufficient evidence was not available to reject the null hypothesis of normality. This confirms the ANOVA normality assumption and the linear regression model for IPMP degradation was generated, see Table H.11.

Table H.9
Main Effects Bench Scale IPMP ANOVA

Terms added sequentially (first to last)	Df	Sum of Sq	Mean Sq	F Value	Pr(F)
UV	**1**	**3066.7**	**3066.7**	**12.44**	**0.002**
Water	**1**	**1137.6**	**1137.6**	**4.61**	**0.046**
H2O2	**2**	**5425.4**	**2712.7**	**11.00**	**0.001**
Time	**1**	**2382.0**	**2382.0**	**9.66**	**0.006**
Residuals	18	4437.8	246.5		

Table H.10
IPMP Bench Scale ANOVA with Interactions

Terms added sequentially (first to last)	Df	Sum of Sq	Mean Sq	F Value	Pr(F)
UV	**1**	**3066.7**	**3066.7**	**17.04**	**0.001**
Water	**1**	**1137.6**	**1137.6**	**6.32**	**0.025**
H2O2	**2**	**5425.4**	**2712.7**	**15.07**	**0.000**
Time	**1**	**2382.0**	**2382.0**	**13.24**	**0.003**
UV lamp:Time	**1**	**1275.1**	**1275.1**	**7.08**	**0.019**
Water:Time	1	76.9	76.9	0.43	0.524
H2O2:Time	2	566.1	283.1	1.57	0.242
Residuals	14	2519.7	180.0		

Table H.11
Linear Regression model for IPMP Removal

	Value	Std. Error	t value	Pr(>|t|)
(Intercept)	-42.8	15.4	-2.79	0.013
UV	**39.4**	**17.5**	**2.26**	**0.037**
Time	**0.1**	**0.0**	**3.61**	**0.002**
Water	**14.0**	**6.5**	**2.14**	**0.047**
H2O22	**25.3**	**8.7**	**2.89**	**0.010**
H2O25	**49.7**	**9.1**	**5.48**	**0.000**
UV lamp:Time	**2.1**	**0.8**	**2.62**	**0.018**

The linear regression model for IPMP is different than the models for MIB and Geosmin. The main difference is that the baseline includes Synthetic water, low UV lamp, 0 mg/L of H_2O_2 and short time. At these low levels an average decrease of 42.8% in IPMP removal is seen. Since it is not possible to have a decrease in percent removal it is acceptable to say that at all low levels of treatment there is no increase in percent removal. The use of Durham water increases IPMP percent removal by 14.0%. Also all treatment variables and the interaction of UV lamp:time in the model were significant. Lastly, the key difference is with time now included as a significant factor the graph contains many experimental conditions, see Figure H.3. Given that time and water are significant factors in IPMP removal it is much harder to make generalizations from the graph since many treatment combinations are possible.

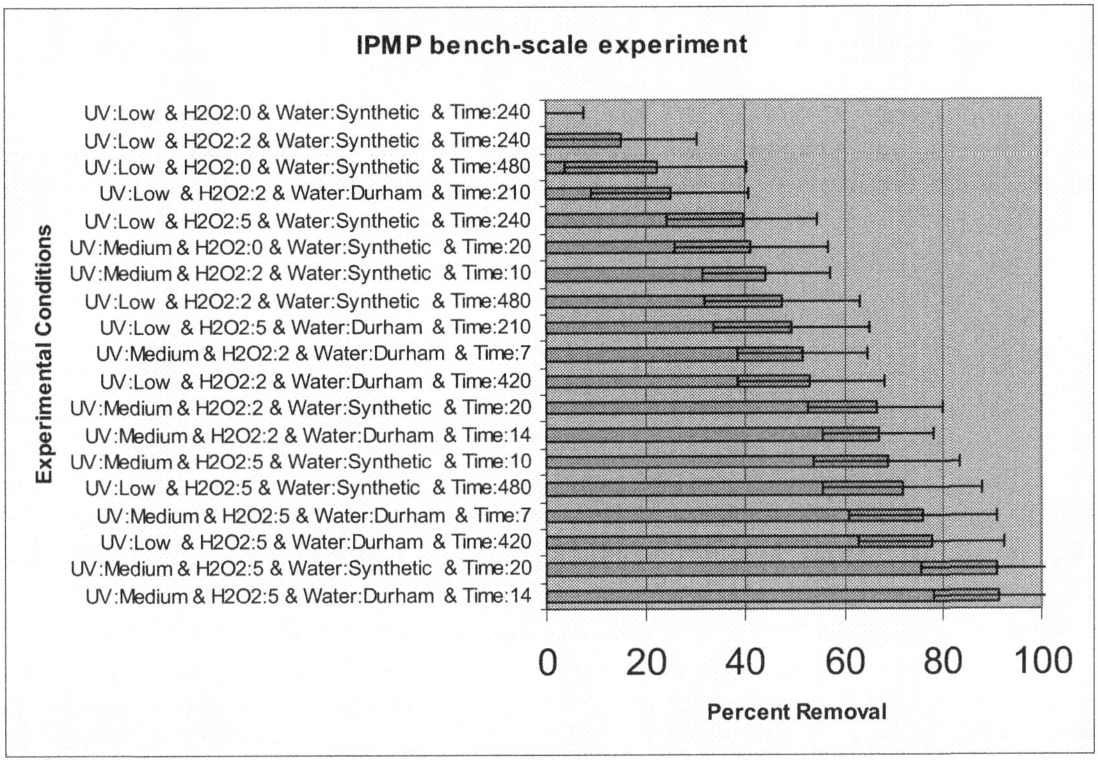

Figure H.3 IPMP Percent Removal according to Linear Regression Model

Pilot Plant

The pilot plant samples consisted of influents, LP (Trojan unit) effluents and MP (Calgon unit) effluents. The influent samples were taken as the concentration of the samples before treatment and the effluents were the concentrations after treatment through the respective reactors. The percent removal of MIB and Geosmin was then calculated by:

[(influent – Trojan effluent)/ influent] * 100 = % removal Trojan Reactor
[(influent – Calgon effluent) / influent] * 100 = % removal Calgon Reactor

The influents were mixed in batches so they could be averaged together while the effluents were taken as individual data. The same analytical procedures that were done in the Bench Scale study were then applied. These included the ANOVA, a validity test for the ANOVA model assumptions, the generation of a linear regression model and a graphical representation of the model.

Only MIB and Geosmin were studied for the pilot plant. The additive linear model for the pilot plant had percent removal as the dependent variable and reactor, UV flow, H_2O_2 dose and contaminant concentration as independent variables. The contaminant concentration was the amount of the taste and odor compound that was spiked into the water. It was either 150 or 300 ng/L of MIB and Geosmin. However, when the samples were analyzed, these concentrations were observed to have great variability and therefore for statistical purposes the observed contaminant concentrations were used, see Table H.2. The basic formula for the main effects of the pilot plant linear model is:

% removal = a + b_1[Reactor] + b_2[UV flow] + b_3[H_2O_2] + b_4[Average Observed Influent Concentration]

The pilot plant was run twice. The first run produced MIB values below the detection limit but showed good Geosmin results. The pilot plant was run again and this time produced MIB results but low Geosmin concentrations. For this reason, the Geosmin data are from the first pilot plant run and the MIB data are from the second pilot plant run. This also explains the difference in H_2O_2 treatment doses between the compounds.

MIB

The factor plots, a box and whisker plot of the treatment variables, were first examined for MIB, see Figure H.4. The horizontal line for each variable represents the median while the upper and lower edges mark the interquartile range. If the error bars among the different variables do not overlap, it can be said that the two treatment variables are different. From the plot it was noted that there is no significant difference between the two different reactors, Trojan and Calgon, and between the two contaminant concentrations 150 and 300 ng/L. This is because the medians are similar and the error bars overlap. However, the factor plot for H_2O_2 dose versus Percent Removal shows that there is a significant difference between 0 and 6 mg/L of H_2O_2. This is because the error bars do not overlap. Nonetheless, the ANOVA needs to be done to confirm the factor plot.

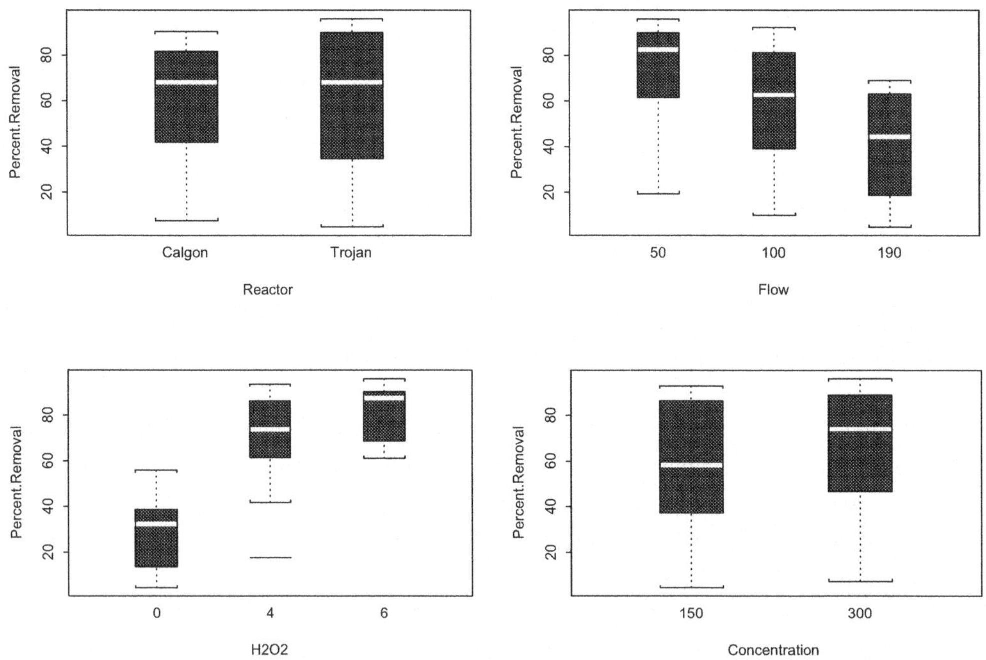

Figure H.4 MIB Factor Plot

The ANOVA was calculated using percent removal as the dependent variable and reactor, UV flow, H_2O_2 dose and average observed influent concentration as independent variables. Only UV flow and H_2O_2 had significant p-values, see Table H.12. The two treatment variables were highly significant with p-values much less than the alpha value of 0.05. Visual inspections were done and the Kolmogorov-Smirnov Test was performed which gave a p-value of 0.5. This does not give sufficient evidence to reject the null hypothesis of normality for the residuals and therefore the linear regression model could be generated, see Table H.13.

The linear regression model further confirms the significance of UV Flow and H_2O_2 as important factors in the removal of MIB. At the low values of 50 gallons per minute UV flow and 0 mg/L of H_2O_2 there is an average 40.9% removal of MIB. However, if the flow is increased a reduction in MIB percent removal is seen. This confirms the hypothesis of a slow UV flow being more effective in contaminant removal due to the extra reaction time between the contaminant and the UV. Also the hypothesis of a higher H_2O_2 dose enhancing compound removal is confirmed. The linear model shows that a change from 0 to 6 mg/L of H_2O_2 adds an additional 54% removal in MIB, which will provide a total average 94.9% MIB removal. These and other treatment combinations can be seen graphically in Figure H.5. The figure also confirms the factor plot's hypothesis of a difference between 0 and 6 mg/L of H_2O_2 since the error bars for these experimental conditions do not overlap.

Table H.12
Pilot Plant MIB ANOVA

	Df	Sum of Sq	Mean Sq	F Value	Pr(F)
Reactor	1	2.7	2.7	0.02	0.879
Flow	**2**	**8880.9**	**4440.4**	**37.87**	**0.000**
H2O2	**2**	**21993.2**	**10996.6**	**93.79**	**0.000**
Avg.MIB.Observed.Influent	1	28.1	28.1	0.24	0.627
Residuals	42	4924.5	117.3		

Table H.13
Linear Regression Model for MIB Removal

	Value	Std. Error	t value	Pr(>|t|)
(Intercept)	40.9	3.3	12.21	0.000
Flow100	**-12.3**	**3.7**	**-3.34**	**0.002**
Flow190	**-31.3**	**3.8**	**-8.22**	**0.000**
H2O24	**38.5**	**3.7**	**10.38**	**0.000**
H2O26	**54.0**	**4.0**	**13.39**	**0.000**

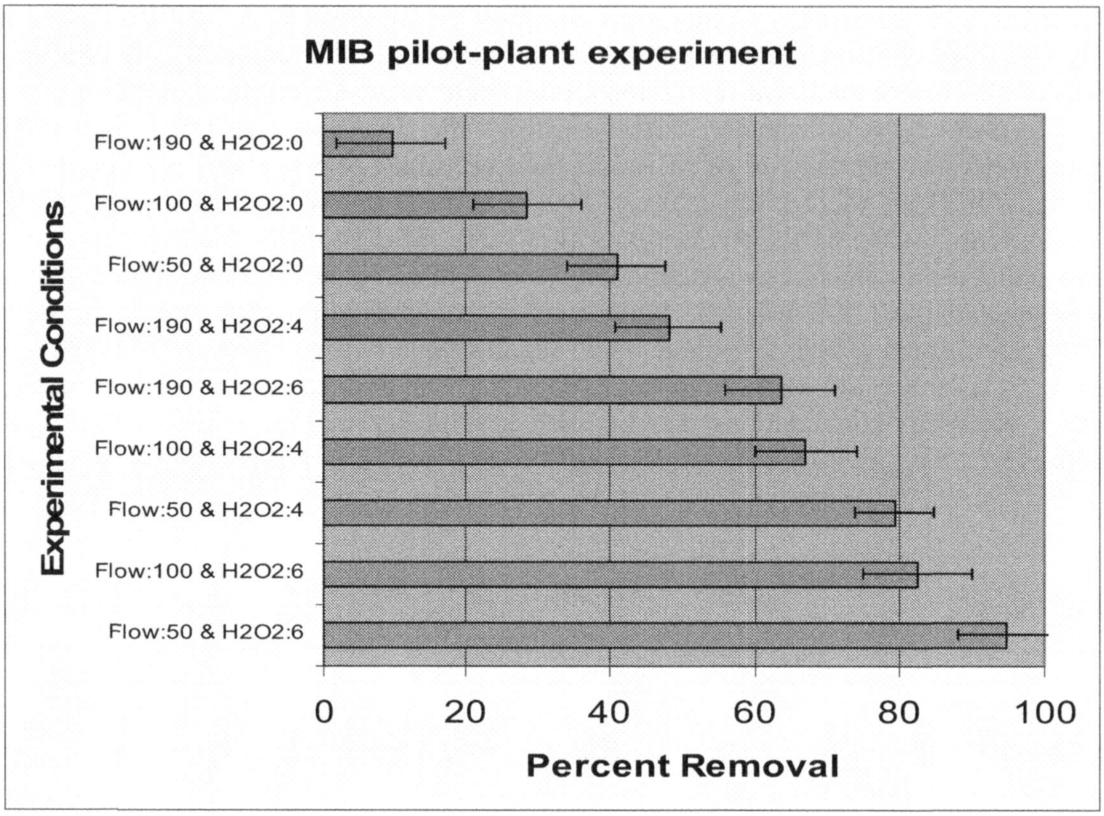

Figure H.5 MIB Percent Removal according to Linear Regression Model

Geosmin

The factor plots were first examined to see which treatment variables would be influential in Geosmin removal, see Figure H.6. From the plots it is evident that there is no difference between the two reactors and concentrations since their medians are similar and the plot's whiskers overlap among the treatment variables. Similarly to MIB, there seems to be a difference between 50 and 190 gallons per minute for flow and among 0 and 7.4 mg/L of H_2O_2. These variables need to be further tested since there are outliers in the flow factor plot and the whiskers overlap for both the flow and H_2O_2 dose factor plots, despite the fact that the medians are unlike.

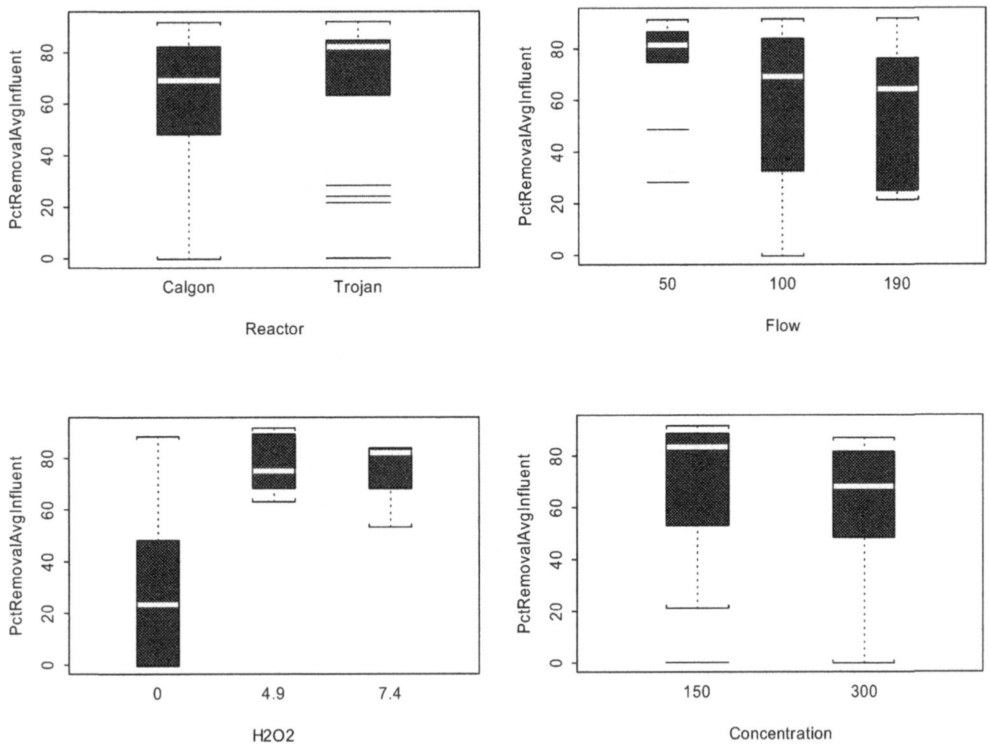

Figure H.6 Geosmin Factor Plot

Before the ANOVA could be calculated the interaction plots of Geosmin were examined and many possible interactions were observed (Data not listed). Therefore, when the ANOVA was determined all possible interactions were included along with the main effects. This model explained 95% of the variability but gave only H_2O_2 dose as significant. As a result, the interactions were removed and just the main effects were taken into consideration. Now H_2O_2 dose and flow were significant and 65% of the variability was explained by the model, see Table H.14.

The diagnostic plots of the residuals were examined and the Kolmogorov-Smirnov test was carried out. The calculated p-value was 0.007 which is much less than the alpha value of 0.05 and therefore sufficient evidence was available to reject the null hypothesis of normality. This indicates that the ANOVA assumption of normality is not met and therefore can not be used. Another model was then analyzed which included only flow and H_2O_2 dose along with the interaction between them, see Table H.15. Only H_2O_2 dose was significant in this ANOVA but the Kolmogorov-Smirnov test now had a p-value of 0.5. The flow:H_2O_2 interaction, which was not significant, had to be included in the ANOVA to achieve normality even if it reduced the flow factor to non-significant. This now met the ANOVA assumption and allowed the linear regression model to be calculated, see Table H.16.

This linear regression model explains 65.3% of the variability. An interesting observation of the model was that even though H_2O_2 dose was significant in the ANOVA only flow was significant in the model. Just at the low values of 50 gallons per minute UV flow and 0 mg/L of H_2O_2 there is on average 55.3% removal of Geosmin. However, when flow was increased to either 100 or 190 gallons per minute there was no effect on the percent removal. When H_2O_2 was increased an increase in total average percent removal was seen. A graphical representation of this model can be seen in Figure H.7. Similar to the bench scale, the Geosmin and MIB models in the pilot plant are very similar.

Table H.14
Pilot Plant Geosmin ANOVA

	Df	Sum of Sq	Mean Sq	F Value	Pr(F)
Reactor	1	258.2	258.2	0.72	0.402
Flow	**2**	**2515.7**	**1257.9**	**3.53**	**0.043**
H2O2	**2**	**14980.8**	**7490.4**	**21.02**	**0.000**
Concentration	1	735.4	735.4	2.06	0.162
Residuals	28	9978.1	356.4		

Table H.15
Pilot Plant Geosmin ANOVA with Flow and H_2O_2

	Df	Sum of Sq	Mean Sq	F Value	Pr(F)
Flow	2	2444.5	1222.3	3.21	0.057
H2O2	**2**	**15150.5**	**7575.2**	**19.92**	**0.000**
Flow:H2O2	4	985.1	246.3	0.65	0.634
Residuals	26	9888.0	380.3		

Table H.16
Linear Regression Model Geosmin Removal

	Value	Std. Error	t value	Pr(>|t|)
(Intercept)	55.3	11.3	4.91	0.000
Flow100	**-33.0**	**14.9**	**-2.22**	**0.036**
Flow190	**-31.9**	**14.9**	**-2.14**	**0.042**
H2O24.9	27.3	14.9	1.83	0.078
H2O27.4	27.8	14.9	1.86	0.074
Flow100H2O24.9	28.9	20.3	1.43	0.166
Flow190H2O24.9	22.4	20.3	1.10	0.280
Flow100H2O27.4	26.2	20.3	1.29	0.209
Flow190H2O27.4	20.3	20.3	1.00	0.328

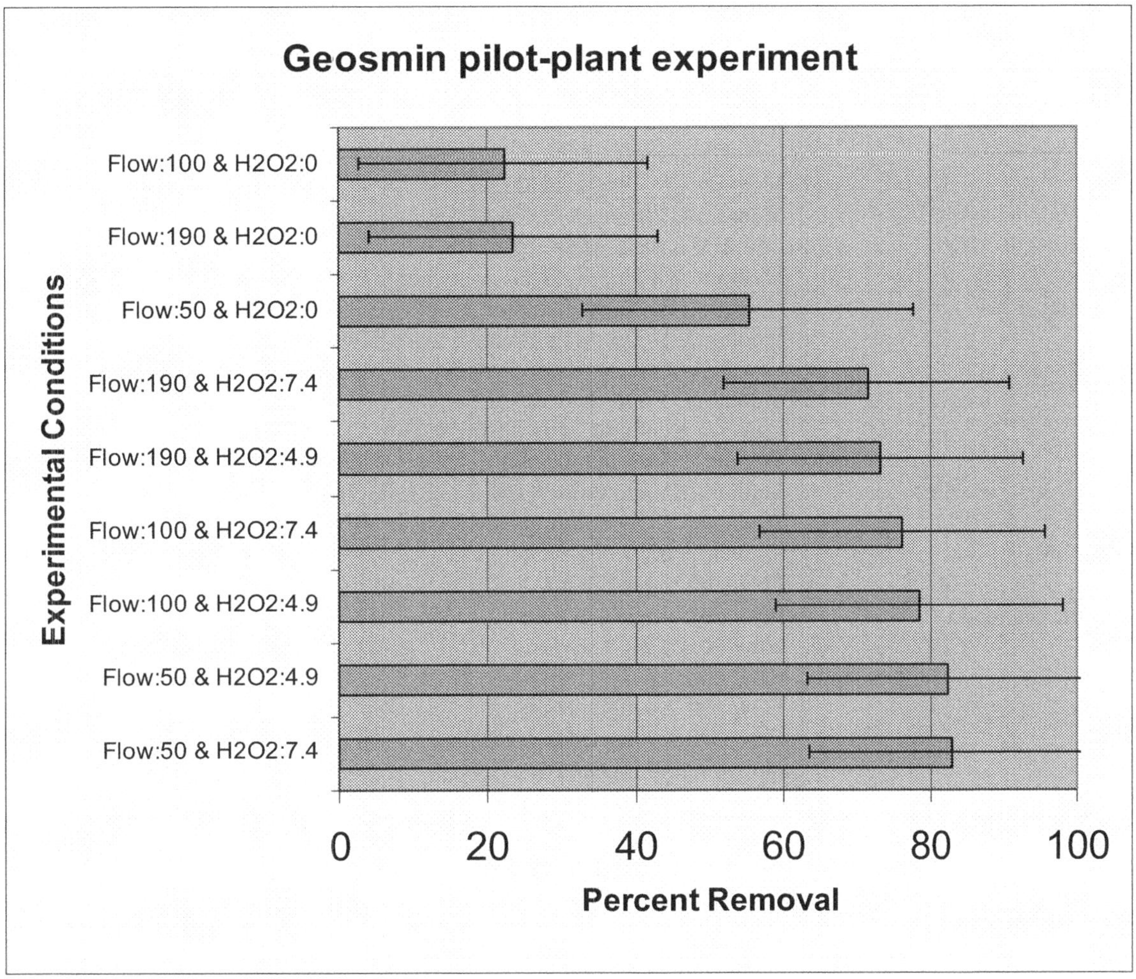

Figure H.7 Geosmin Percent Removal According to the Linear Regression Model

Discussion

The statistically significant treatment combinations for the removal of taste and odor compounds at the bench scale and pilot plant could now be examined. The primary and secondary significant combinations were selected by examining the graphical representations of the linear models. The primary treatment combinations were found by grouping all the experimental conditions whose error bars overlapped with the condition at the highest percent removal, see Table H.17. Similarly the secondary treatment combinations were found by grouping from the next available experimental condition, see Table H.18.

Table H.17
Primary Significant Treatment Combinations

	Primary Combinations
Bench Scale: MIB	Medium UV 5 mg/L H_2O_2 Low UV 5 mg/L H_2O_2 Medium UV 2 mg/L H_2O_2
Bench Scale: Geosmin	Medium UV 5 mg/L H_2O_2 Low UV 5 mg/L H_2O_2 Medium UV 2 mg/L H_2O_2
Bench Scale: IPMP	Medium UV Durham Water 5 mg/L H_2O_2 Time 14 Medium UV Synthetic Water 5 mg/L H_2O_2 Time 20 Low UV Durham Water 5 mg/L H_2O_2 Time 420 Medium UV Durham Water 5 mg/L H_2O_2 Time 7 Low UV Synthetic Water 5 mg/L H_2O_2 Time 480 Medium UV Synthetic Water 5 mg/L H_2O_2 Time 10 Medium UV Durham Water 2 mg/L H_2O_2 Time 14 Medium UV Synthetic Water 2 mg/L H_2O_2 Time 20
Pilot Plant: MIB	Flow 50 6 mg/L H_2O_2 Flow 100 6 mg/L H_2O_2
Pilot Plant: Geosmin	Flow 50 7.4 mg/L H_2O_2 Flow 50 4.9 mg/L H_2O_2 Flow 100 4.9 mg/L H_2O_2 Flow 100 7.4 mg/L H_2O_2 Flow 190 4.9 mg/L H_2O_2 Flow 190 7.4 mg/L H_2O_2 Flow 50 0 mg/L H_2O_2

Table H.18
Secondary Significant Treatment Combinations

	Secondary Combinations
Bench Scale: MIB	Low UV 2 mg/L H_2O_2 Medium UV 0 mg/L H_2O_2 Low UV 0 mg/L H_2O_2
Bench Scale: Geosmin	Low UV 2 mg/L H_2O_2 Medium UV 0 mg/L H_2O_2 Low UV 0 mg/L H_2O_2
Bench Scale: IPMP	Low UV Durham Water 2 mg/L H_2O_2 Time 420 Medium UV Durham Water 2 mg/L H_2O_2 Time 7 Low UV Durham Water 5 mg/L H_2O_2 Time 210 Low UV Synthetic Water 2 mg/L H_2O_2 Time 480 Medium UV Synthetic Water 2 mg/L H_2O_2 Time 10 Medium UV Synthetic Water 0 mg/L H_2O_2 Time 20 Low UV Synthetic Water 5 mg/L H_2O_2 Time 240 Low UV Durham Water 2 mg/L H_2O_2 Time 210 Low UV Synthetic Water 0 mg/L H_2O_2 Time 480
Pilot Plant: MIB	Flow 50 4 mg/L H_2O_2 Flow 100 4 mg/L H_2O_2
Pilot Plant: Geosmin	Flow 190 0 mg/L H_2O_2 Flow 100 0 mg/L H_2O_2

Bench Scale

From the tables it is evident that at the bench scale, MIB and Geosmin have the same primary and secondary significant combinations for compound removal. The primary combinations consist of a medium/low UV lamp and 5 mg/L of H_2O_2 and a medium UV lamp and 2 mg/L of H_2O_2. This supports the hypothesis of a high H_2O_2 dose being important in taste and odor compound removal since the additional H_2O_2 generates more hydrogen radicals. The p-values for 5 mg/L of H_2O_2 verify the importance of H_2O_2 dose as a treatment factor in taste and odor compound removal. Also from the graphical representations of the linear models, a higher percentage of Geosmin is removed. Just at the low significant baseline values of low UV lamp and 0 mg/L H_2O_2 there is a 17.5% removal of Geosmin and 11.1% removal of MIB on average. This also supports the hypothesis of Geosmin being easier to degrade due to less steric hindrance. However, the error bars on the graphical linear model cover a large range and therefore this hypothesis can not be proved.

IPMP is a more complicated compound to examine since time is a significant factor and analysis was only conducted at the bench scale experiment. Many primary and secondary interactions existed since all four factors and an interaction were significant, see Table H.17 and H.18. At the low levels there was no increase in percent removal. However, the linear regression model indicates that a 5 mg/L dose of H_2O_2 and a medium pressure UV lamp enhance IPMP degradation. This again supports the hypothesis of a high H_2O_2 dose being of extreme importance in compound removal. Also the experimental conditions with 2 and 5 mg/L of H_2O_2 removed more IPMP on average but there was a condition that was listed as a secondary combination which included 0 mg/L of H_2O_2. However, it was paired with a medium pressure UV lamp and was run at a long time for its batch, 20 seconds. It is likely the medium pressure UV lamp and the long exposure time helped to increase its average percent removal rate. Time was challenging to interpret since the short times were coupled with the medium pressure UV and the longer times were paired with a low pressure UV. Overall, the graph illustrates that a longer exposure time does have a higher average percent removal rate, see Figure H.3. Again, this can not be statistically differentiated since the error bars cover a large area.

It is important to verify that the observed best removed samples are similar to the linear model results. From Table H.2 it is seen that the best removed samples at the bench scale consist of a medium UV lamp and 5 mg/L of H_2O_2 or with a low UV lamp and 5 mg/L of H_2O_2. This is verified by the primary significant treatment combinations that were obtained from the linear model graphs of MIB and Geosmin at the bench scale. Therefore, the observed bench scale results are represented well by the linear models.

The significant treatment factors for each compound at the bench scale can be easily seen in Figures H.8-H.10. The graphs can be interpreted by taking 0% as the average percent removal and the treatment factors either increase or decrease the total average percent removal. It can be seen from the graphs that a 5 mg/L H_2O_2 dose enhances the average percent removal for all three compounds. IPMP shows that changing the UV to medium also increases IPMP average percent removal. Striking is the fact that only IPMP shows an effect of the background water quality.

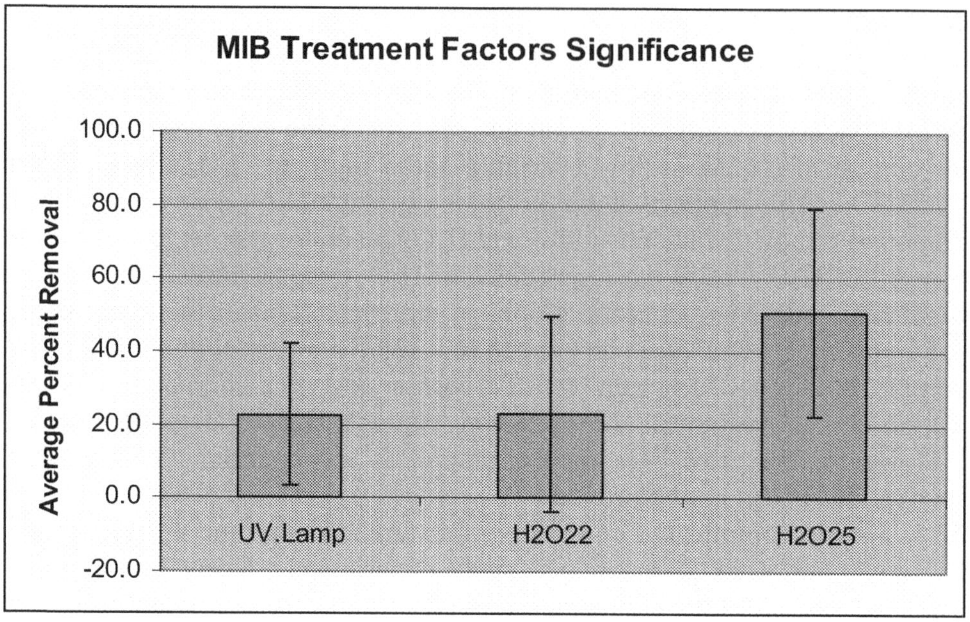

Figure H.8 MIB Treatment Factor Effects (H_20_22 = 2 mg/L & H_20_25 = 5 mg/L)

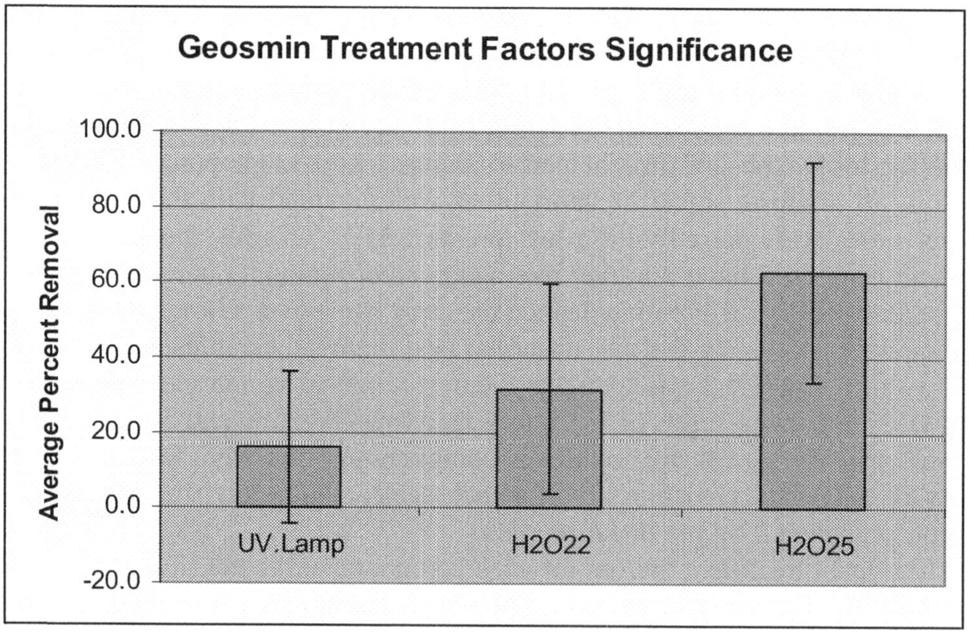

Figure H.9 Geosmin Treatment Factor Effects (H_20_22= 2 mg/L & H_20_25= 5 mg/L)

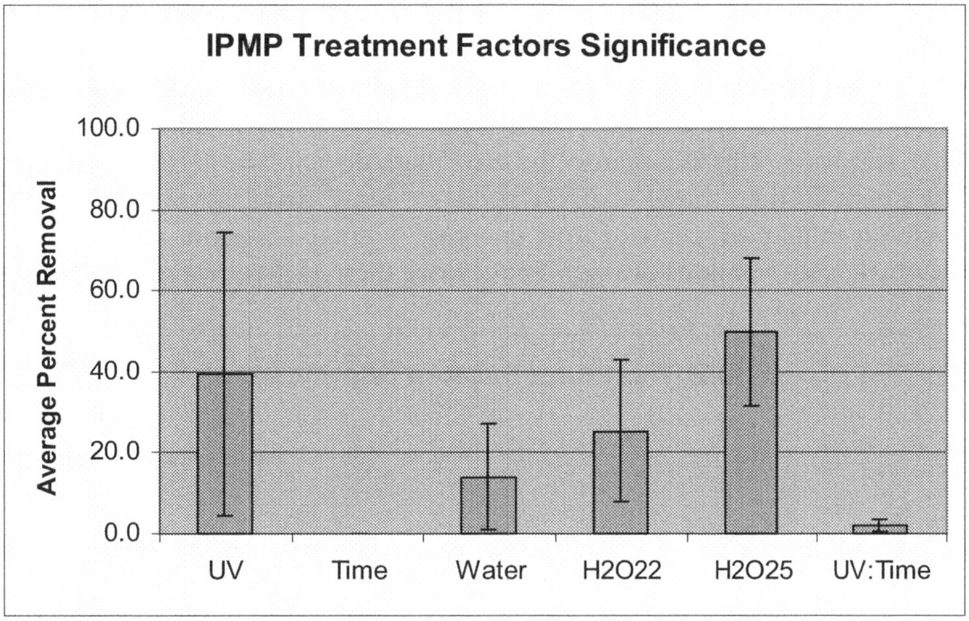

Figure H.10 IPMP Treatment Factor Effects (H_20_22= 2 mg/L & H_20_25= 5 mg/L)

Pilot Plant

Table H.17 and H.18 show MIB and Geosmin had only UV flow and H_2O_2 as significant treatment factors in their linear models. The experimental conditions with 0 mg/L of H_2O_2 had the lowest average percent removal, with the slow UV flow removing more, while the highest percent removal was seen with the slow UV flow and high H_2O_2 dose. This seems to be accurate since a low UV flow would allow more time for the UV to react while a high UV flow would pass through the reactor quickly thereby allowing less time for reactions to occur. Also the low UV flow would be able to react with the high H_2O_2 dose to produce more hydroxyl radicals which will further help in degrading MIB and Geosmin.

Even though both MIB and Geosmin show similar graphical representations of the linear models some differences do exist. The main difference is that the error bars on the MIB graph are small and do not overlap much between experimental conditions while the error bars on Geosmin cover a larger range, see Figure H.5 and H.7. This is the reason for Geosmin having many primary combinations for percent removal, see Table H.17. The error bars overlap among various experimental conditions and therefore more combinations are included as primary combinations. Also since the Geosmin data were taken from the first pilot plant run and MIB from the second pilot plant run their H_2O_2 treatment doses differ slightly, see Figure H.4 and H.6. An interesting difference is that the UV flow gave a negative removal rate for Geosmin, which means that increasing the flow causes a decrease of Geosmin removal. Lastly, at the low levels of 0 mg/L of H_2O_2 and 50 gallons per minute UV flow, 40.9% of MIB and 55.3% of Geosmin is removed on average. This again illustrates that Geosmin should be removed at a higher rate than MIB but it can not be verified statistically due to the large error bars for Geosmin.

Similar to the bench scale experiment, verification needs to be shown that the best removed samples are similar to the statistical results. The best removed samples at the pilot plant were 100 gpm flow and 6 mg/L of H_2O_2, 50/100 gpm flow and 4.9 mg/L of H_2O_2 and 50/100 gpm flow and 7.4 mg/L of H_2O_2, see Table H.3 and H.4. These again are seen as the primary significant treatment combinations for MIB and Geosmin at the pilot plant. Therefore, the observed percent removals from Tables H.2-4 are good representations of the linear models.

The treatment factor effects on the average percent removal at the pilot plant can be seen in Figure H.11 and H.12. It can be clearly seen that a 6 mg/L dose of H_2O_2 significantly enhances MIB removal. This is because the error bars do not overlap with the other treatment factors. The graph further shows that all levels of flow decrease average percent removal. Geosmin is similar in that all levels of flow decrease average Geosmin removal but it is hard to see which treatment combinations are significant since there is great variability.

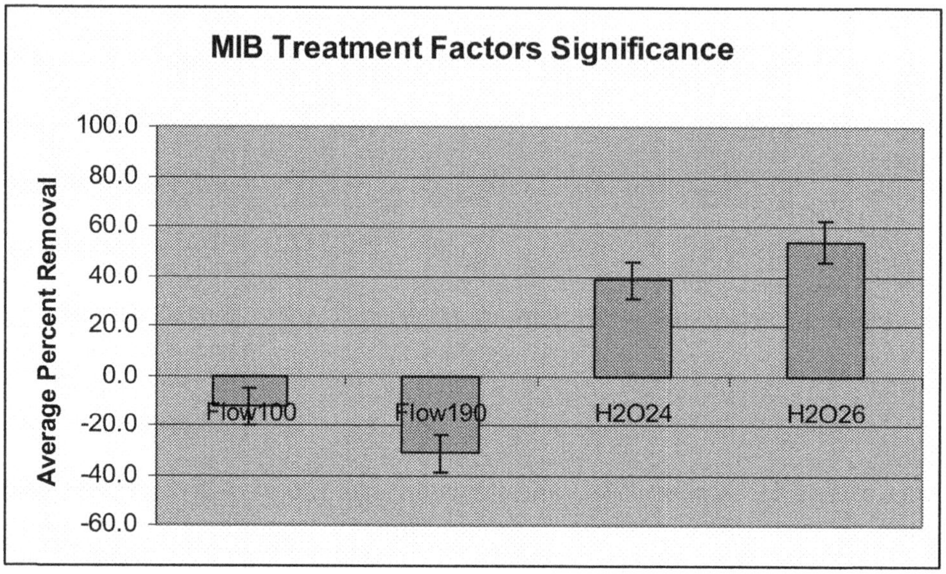

Figure H.11 MIB Treatment Factor Effects (H_20_24= 4 mg/L & H_20_26= 6 mg/L)

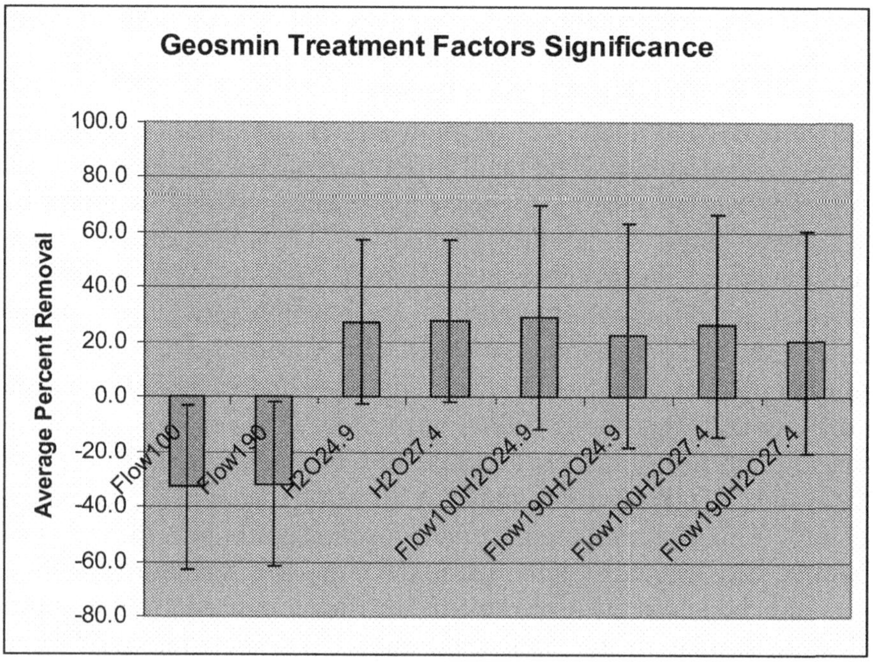

Figure H.12 Geosmin Treatment Factor Effects ($H_20_24.9$ =4.9 mg/L & $H_20_27.4$ =7.4 mg/L)

CONCLUSION

Similar treatment factors were statistically analyzed for the removal of taste and odor compounds in both a bench scale and pilot plant experiment. From the linear regression models it is evident that a high H_2O_2 dose, a medium pressure UV lamp and a low UV flow enhance the removal of MIB, Geosmin and IPMP. Time seems to be a secondary factor where a longer time of exposure increases the average removal rate. The best combinations for removal consists of a either a medium pressure UV lamp and high H_2O_2 dose or a low UV flow and high H_2O_2 dose. Furthermore, the bench scale seemed to be a good representation of the pilot plant.

These results agree with the hypotheses of a combination of UV and H_2O_2 being critical and a combination of high H_2O_2 dose, low UV flow and longer exposure time enhancing compound degradation. Geosmin also showed a higher percent removal than MIB on average, which was hypothesized due to less steric hindrance. IPMP needs some form of treatment to assist in compound removal since at low experimental factor values it had a supposedly negative average percent removal.

Note that these are initial results based on the data set available and to verify the significance of the treatment factors another study should be conducted with a larger sample size.

REFERENCES

Agriculture Canada. 1982 *Guide to the chemicals used in crop protection, 7th edition.* Agriculture Canada Publication No. 1093.

Andrews, S.A.; P.M. Huck, A.J. Chute, J.R. Bolton, W.A. Anderson. 1995. UV Oxidation for Drinking Water - Feasibility Studies for Addressing Specific Water Quality Issues. *Proc. AWWA WQTC*, New Orleans, LA. 1881-1898.

Andrews, S.A., S. Peldszus, C. Moralejo, W.A. Anderson, P.M. Huck. 1999. Stable By-Products of the Photo-Oxidative Advanced Oxidation of BTEX Compounds in Groundwater. *Proc. AWWA WQTC*, November, Tampa, FL

APHA, AWWA, and WEF (American Public Health Association, American Water Works Association, and Water Environment Federation). 1998. *Standard Methods for the Examination of Water and Wastewater.* 20th ed. Washington, D.C.: APHA

APHA, AWWA, and WEF (American Public Health Association, American Water Works Association, and Water Environment Federation). 2000. *Standard Methods for the Examination of Water and Wastewater.* 21st Ed., Washington D.C.: APHA

ATSDR (Agency for Toxic Substances and Disease Registry). 1989. Toxicological profile for N-nitrosodimethylamine. Atlanta, GA: U.S. Department of Health and Human Services, Public Health Service.

ATSDR (Agency for Toxic Substances and Disease Registry) 1996. Toxicological profile for Diazinon" Atlanta, GA: U.S. Department of Health and Human Services, Public Health Service.

ATSDR (Agency for Toxic Substances and Disease Registry). 2000. Toxicological profile for Toluene" Atlanta, GA: U.S. Department of Health and Human Services, Public Health Service.

AWWA (American Water Works Association). 2003. *Mainstream*, pg. 3, March.

AwwaRF (Awwa Research Foundation). 1987. Identification and Treatment of Taste and Odor in Drinking Water, Ed. I.H. Suffet & J. Mallevialle, Awwa Research Foundation, American Water Works Association, Denver CO, ISBN #0-89867-392-5.

AwwaRF (Awwa Research Foundation) and Lyonnaise des Eaux. 1995. Advances in Taste-and-Odor Treatment and Control, Ed. I.H. Suffet, J. Mallevialle, and E. Kawczynski. AwwaRF and AWWA. Denver, CO.

Baxendale, J. H.; Wilson, J. A. 1957. *Trans. Faraday Soc.*, *53*, 344-356.

Beltran, F. J., G. Ovejero, B. Acedo. 1993. *Wat. Res.*, *27*, 1013-1021.

Beltran, F. J., M. Gonzalez, J.F. Rivas, P. Alvarez. 1996. *Environ. Toxicol. Chem.*, *15*, 868-872.

Bolton, J.R. and M.I. Stefan. 2002. Fundamental photochemical approach to the concepts of fluence (UV dose) and electrical energy efficiency in photochemical degradation reactions. *Res. Chem. Intermed.* 28(7-9): 857-870.

Bolton, J.R., S.R. Fitzsimmons, L.J. Yoo, Y.F. Shen. 2001. Ultraviolet Treatment and Potential Formation of N-Nitrosodimethylamine (NDMA) in a Water Reclamation Treatment Plant. *Proceedings International Ultraviolet Association*, 1st Annual Congress, Washington, DC.

Cairns, W. 1996. Proceedings, Disinfection Specialty Conference, Water Environment Federation, Portland, OR.

Chan, G.Y S., M.J. Hudson, N.S. Isaacs. 1992. *J. Phys. Org. Chem.*, *5*, 600-608.

Cooper, D. and L. Keller. 1993. Advanced oxidation technology demonstration at Ohsweken Six Nations Indian Reserve. *Proc. - Tech. Semin. Chem. Spills 10th.,* Environ. Can., Environ. Prot., Technol. Dev. Branch, Ottawa, Canada; pp. 51-63.

Doong, R., W-H Chang. 1998. *Chemosphere*, *37*, 2563-2572.

Dunn, J., T. Ott, W. Clark. 1995. "Pulsed light treatment of food and packaging". *Food Technol.*, 49, 95

Ecobichon, D.J., 1991. in *Casarett and Doull's Toxicology: The Basic Science of Poisons, 4th Ed.*; M.O. Amdur, J. Doull, C.D. Klassen (Ed.'s); Pergamon Press, New York, 1991, pp. 565-622.

Environment Canada. 2001. Priority substances list assessment report: N-nitrosodimethylamine (NDMA) (Environment Canada catalog number En40-215/53E), Co-published by Health Canada.

EXTOXNET. 2003. http://pmep.cce.cornell.edu/profiles/extoxnet/carbaryl-dicrotophos/carbaryl-ext.html#13

Ferguson, D.W., M.J. McGuire, B. Koch, R.L. Wolfe, E.M. Aieta, 1990. *Journal AWWA*, 82 (4), 181-191.

Frimmel, F.H., D.P. Hessler, 1994. in *Aquatic and Surface Photochemistry*; Helz, G. R., Zepp, R. G., Crosby, D. G., Eds.; CRC Press: Boca Raton; pp 137-147.

Geiger, G., J.R. Huber. 1981. *Helv. Chim. Acta*, *64*, 989-995.

Gerecke A.C., D.L. Sedlak. 2003. *Environ. Sci. Technol., 37(7)*, 1331-1336.

Glaze, W.H., R. Schep, W. Chauncey, E.C. Ruth, J.J. Zarnoch. 1990. Evaluating Oxidants for the Removal of Model Taste and Odor Compounds from a Municipal Water Supply. *Journal AWWA*, 82:5:79.

Glod, D., Z. Pawlak. 1983. *Stud. Mater. Oceanol.*, 41, 181-192.

Griffini, O., M.L. Bao, D. Burrini, D. Santianni, C. Barbieri, F. Pantani. 1999. *J. WSRT-AQUA, 48(5)*, 177-185.

Hattori, K. 1988. *Wat. Sci, Technol.*, 20(8/9):237-244.

Hayes T.B., A. Collins, M. Lee, M. Mendoza, N. Noriega, A.A. Stuart, A. Vonk. 2002. *Proc. Natl. Acad. Sci. USA, 99 (8)*, 5476-5480.

Hayes T., K. Haston, M. Tsui, A. Hoang, C. Haeffele, A. Vonk. 2003. *Environ. Health Persp. 111 (4)*, 568-575.

Hessler, D. P., V. Gorenflo, F.H. Frimmel. 1993. *Acta Hydrochim. Hydrobiol., 21*, 209-214.

Ho, T-F. L., J.R. Bolton, E. Lipczynska-Kochany. 1996. *J. Adv. Oxid. Technol., 1*, 170-178.

Hrudey, S.E., P.M. Huck, M.J. Mitton, S.L. Kenefick. 1995. *Wat. Sci. Tech.*, 31(11), 195-201.

Hu, J.Y.; T. Aizawa, Y. Magara. 1997. *Water Sci. Technol., 35(7)*, 219-226.

Huck, P.M.; W.B. Anderson, C.L. Lang, W.A. Anderson, J.C. Fraser, S.Y. Jasim, S.A. Andrews, G. Pereira. 1995. Ozone vs PEROXONE for Geosmin and 2-Methylisoborneol Control: Laboratory, Pilot and Modeling Studies. P*roceedings, AWWA Annual Conference and Exhibition.*

Jobb, D.B., R.B. Hunsinger, O. Meresz, and V. Taguchi. 1994. Ultraviolet degradation of N-nitrosodimethylamine (NDMA) in the Ohsweken (Six Nations) water supply. P*roceedings, AWWA Annual Conference and Exhibition,* pp. 211.218.

Kang, J-W, K-H. Lee. 1997. *Environ. Eng. Sci.*, 14(3), 183-192.

Kearney, P.C., M.T. Muldoon, C.J. Somich. 1987. *Chemosphere, 16*, 2321-2330.

Kim, J.H., S.E. Feagley. 1998. *J. Environ. Sci. Health, Pt. B: Pestic., Food Contam., Agric. Wastes, 33(5)*, 529-546.

Klassen, N.V., D. Marchington, H.C.E. McGowan. 1994. H_2O_2 Determination by the I^{3-} Method and by $KMnO_4$ Titration. *Anal. Chem.*, 66(18): 2921-2925.

Kleiser G., F.H. Frimmel. 2000. Removal of precursors for disinfection by-products (DBPs) - differences between ozone- and OH-radical-induced oxidation. *Sci. Total Environ.*, 256 (1): 1-9.

Kobylinski, E.A. and B.W. Peterman. 1979. Evaluation of ozone oxidation and UV degradation of dimethylnitrosamine. Fort Detrick, Frederick, MD 21701, U.S. Army Medical research and Development Command; pp. 1-16.

Koch, B., J.T. Gramith, M.S. Dale, D.W. Ferguson. 1992. *Wat. Sci. Tech.*, 25 (2), 291-298.

Kochany, J. and R.J. Maguire. 1994. *J. Agric. Food Chem., 42*, 406-412.

Kruithof, J.C., P.C. Kamp, M. Belosevic. 2002. *Water Supply, 2*, 113–122.

Ku, Y., J-L. Chang, S-C. Cheng. 1998. *Wat. Air and Soil Poll., 108*, 445-456.

Kwan A. et al. 1996. Proceedings, Disinfection Specialty Conference, Water Environment Federation, Portland, OR.

Lalezary, S., M. Pirbazari, M.J. McGuire. 1986. Oxidation of Five Earthy-Musty Taste and Odor Compounds. *Journal AWWA*, 78:3:62-69.

Lang, C.L., P.M. Huck, W.A. Anderson, S.A. Andrews, 1995. Experimental Verification of a Simplified Kinetic Model for Removing Geosmin and MIB Using Ozone/Peroxide *Proc. American Water Works Association Water Quality Technology Conference*, New Orleans, Louisiana, November.

Lin, Y.-J., K. Makesh, A. Shaw, A. G. Gupta. 1999. *Ecotoxicol. Environ. Saf.*, 43(1), 35-37.

Liu, W., S.A. Andrews, M.I. Stefan, J.R. Bolton. 2003. Optimal Methods for Quenching H_2O_2 Residuals Prior to UFC Testing. *Wat. Res.*, 37, 3697-3703.

Liu, W., S.A. Andrews, J.R. Bolton, K.G. Linden, C. Sharpless, and M. Stefan. 2002a. Comparison of disinfection byproduct (DBP) formation from different UV technologies at bench scale; WST: Water Supply 2 (5&6): 515-521; IWA Publishing.

Liu, W., Andrews, S.A., Sharpless, C., Stefan, M., Linden, K. G., and Bolton, J.R. 2002b. Bench-Scale Investigations into Comparative Evaluation of DBP Formation from Different UV/H_2O_2 Technologies, *2002 AWWA Water Quality Technology Conference Proceedings*, AWWA Water Quality Technology Conference, Seattle, WA, November, CD ROM.

Mabury, S.A. and D.G. Crosby. 1996. *J. Agric. Food Chem., 44(7)*, 1920 -1924.

Malley, J.P., J.P. Shaw, J.R. Ropp. 1995. Evaluation of By-Products Produced by Treatment of Groundwaters with Ultraviolet Irradiation. AWWARF and AWWA, Denver, CO.

Mansour, M. 1985. *Bull. Environ. Contam. Toxicol.*, 34, 89-95.

Mathew, R. and S.U. Khan. 1996. *J. Agric. Food Chem.*, *44*, 3996-4000.

Morioka, T., N. Motoyama, H. Hoshikawa, A. Murakami, M. Okada, T. Moniwa. 1993. *Ozone Sci. Eng.*, 15, 1-18.

Morowitz, H.J. 1950. Absorption Effects in Volume Irradiation of Microorganisms. *Science*, 111:229.

Murov, S.L., I. Carmichael, G.L. Hug. 1993. *Handbook of photochemistry, 2nd ed.*; Marcel Dekker, Inc.: New York; pp 330-336.

NDRL (Notre Dame Radiation Laboratory). 2003. *Database of ·OH reaction rate constants* (http://www.rcdc.nd.edu/compilations/Hydroxyl/OH_576.HTM).

Nick, K. and H.F. Schoeler. 1995. *Vom Wasser*, *84*, 271-286.

Peldszus, S., P.M. Huck, S.A. Andrews. 1996. Determination of short-chain aliphatic, oxo- and hydroxy-acids in drinking water at low microgram per liter concentrations. J. Chromatography A. 723: 27-34

Persson, E. 1988. *Wat. Sci. Technol.*, 20 (8/9), 283-288.

Peyton, G.R.; Bell, O.J.; Girin, E.; LeFaivre, M.H.; Sanders, J.1998. Effect of bicarbonate alkalinity on performance of advanced oxidation processes. AWWARF and AWWA, Denver, CO.

Pignatello, J.J. Y. Sun, 1995. *Wat. Res.*, *29*, 1837-1844.

Rahn R.O. 1997. *Photochem. Photobiol.*, *66*, 450-455.

Rehberg, K.; Mannschott, P.; Erdinger, L.; Sonntag, H. G.; Schramm, B. 1994. *Vom Wasser*, *83*, 139-156.

Schwarzenbach, R.P., P.M. Gschwend, D.M. Imboden. 1993. *Environmental Organic Chemistry, 1st Ed.*; Wiley-Interscience, New York, 681 pp.

Sharpless, C.M. C.I. Chou, A.A. Mofidi, K.G. Linden, 2001. *Proceedings Water Quality and Technology Conference*, American Water Works Association, Nashville, TN.

Sharpless, C.M. and Linden, K.G. 2003. Experimental and Model Comparisons of Low- and Medium-Pressure Hg Lamps for the Direct and H_2O_2 Assisted UV Photodegradation of N-nitrosodimethylamine in Simulated Drinking Water. *Environ. Sci. Technol.*, Vol. 37 No. 9, pp. 1933-1940

Sharpless, C.M., Seibold, D.A., and Linden, K.G. 2003. Nitrate photosensitized degradation of atrazine during UV water treatment. *Aquat. Sci.* Vol. 65, No. 4, pp. 359-366

Speitel, G.E. and J.M. Symons. 1999. Advanced Oxidation/Biodegradation Processes for the Destruction of TOC and Disinfection By-Product Precursors, Final Report, Project #289, Awwa Research Foundation and Awwa, Denver CO.

Stefan, M.I., J.R. Bolton, S.R. Cater. 1999. UV light-driven degradation of N-nitrosodimethylamine. *Book of Abstracts*, The 5th Intl. Conf. on Advanced Oxidation Technologies for Water Remediation, May 24-28, 1999, Albuquerque, NM; p. 109.

Stefan, M.I. and J.R. Bolton. 2002. *Helv. Chim. Acta, 85(5)*, 1416-1426.

Stefan, M.I., C.M. Sharpless, K.G. Linden, J.R. Bolton. 2001. *Proceedings Water Quality and Technology Conference*, American Water Works Association, Nashville, TN.

Stefan, M.I., M. Siddiqui, K. Atasi, K.G. Linden. 2002. *Proceedings Water Quality and Technology Conference*, American Water Works Association, Seattle, WA.

Summers, R.S., S.M. Hooper, G. Solarik, D. Owen. 1996. Assessing DBP Yield: Uniform Formation Conditions. *J.AWWA*, 81(7), 80-93.

Sundstrom, D.W., B.A. Weir, H.E. Klei. 1989. *Environ. Prog.*, 8(1), 6-11.

U.S. EPA. 2002. *2002 Edition of the Drinking Water Standards and Health Advisories* (EPA 822-R-02-038); Office of Water, U.S. Environmental Protection Agency, Washington, DC.

USGS (United States Geological Survey). 1999. *The Quality of Our Nation's Waters: Nutrients and Pesticides: U.S. Geological Survey Circular 1225*, 82 pp.; U.S. Geological Survey, Washington, D.C.

Wan, H.B., M.K. Wong, C.Y. Mok, 1994. *J. Agric. Food Chem.*, *42*, 2625-2630.

Wilson, R.I., S.A. Mabury, 2000. *J. Agric. Food Chem.*, 48, 944-950.

Wink, D.A., R.W. Nims, M.F. Desrosiers, P.C. Ford, L.K. Keefer. 1991. *Chem. Res. Toxicol. 4*, 510-512.

Wolfe, N.L., R.G. Zepp, D.F. Paris. 1978. *Wat. Res.*, 12, 565-571.

Yager, J.E. and C.D. Yue. 1988. *Environ. Toxicol. Chem.*, 7, 1003-1011.

Zheng, M., S.A. Andrews, J.R. Bolton, 1999a. Impacts Medium Pressure UV and UV/H_2O_2 Treatments on Disinfection Byproduct Formation. Proceedings, *American Water Works Association Annual Conference and Exhibition*, Chicago, Illinois, June.

Zheng, M., S.A. Andrews, J.R. Bolton, 1999b. Impacts Medium Pressure UV and UV/H_2O_2 on Subsequent THM and HAA Formation. Proceedings, *American Water Works Association Water Quality Technology Conference*, Tampa, Florida, November.

ABBREVIATIONS AND SYMBOLS

ABBREVIATION	DEFINITION
ANOVA	analysis of variance
AOP	advanced oxidation process
AOT	advanced oxidation technology
ATZ	atrazine
AwwaRF	Awwa Research Foundation
BPI	Bolton Photosciences, Incorporated
cm	centimeter(s)
D	Duke University
DBP(s)	disinfection byproduct(s)
D/DBPR	disinfectants and disinfection byproducts rule
DWSD	Detroit Water and Sewerage Department
E	Irradiance (mW/cm^2)
E_{EO}	electrical energy per order
EED	electrical energy dose
g	gram
GC/MS	gas chromatograph/mass spectrometer
gpm	gallon(s) per minute
GRW	Grand River Water
GSM	geosmin
$H_{200\text{-}300}$	UV dose from 200 to 300 nm (mJ/cm^2)
H_2O_2	hydrogen peroxide
HAA(s)	haloacetic acid(s)
hr(s)	hour(s)
HPLC	high performance liquid chromatography
IPMP	2-isopropyl-3-methoxypyrazine
IUPAC	International Union of Pure and Applied Chemists
J	joule(s)
kW•hr	kilowatt-hour(s)
L	liter(s)
LC	liquid chromatography
LP	low-pressure
LPHO	low-pressure high output
LT2ESWTR	Long-Term 2 Enhanced Surface Water Treatment Rule

µm	micrometer(s)
µs	microsecond
µW	microwatt(s)
m	meter(s)
MCL	maximum contaminant level
mg/L	milligram(s) per liter
mgd, MGD	million gallon(s) per day
MIB	2-methylisoborneol
min	minutes
mJ	millijoule(s)
mL	milliliter(s)
mm	millimiter(s)
mM	millimolar
MP	medium-pressure
mW	milliwatt(s)
MWD	Metropolitan Water District
MWTP	Mannheim Water Treatment Plant
n	number of samples
NDMA	N-nitrosodimethylamine
NOM	natural organic matter
nm	nanometer(s)
ntu, NTU	nephelometric turbidity unit(s)
OH	hydroxyl radical
pH	negative logarithm of the effective hydrogen-ion concentration
ppm	parts per million
psi	pounds per square inch
P-UV	Pulsed UV
Q	flow
QA/QC	quality assurance / quality control
R	correlation coefficient
s	second(s)
SDWA	Safe Drinking Water Act
SPME	solid phase micro-extraction
sq	square(s)
SW	synthetic water
t	time
T	temperature
THM(s)	trihalomethane(s)

TOC	total organic carbon
TRC	total residual chlorine
UCLA	University of California at Los Angeles
UFC	uniform formation conditions
U.S.	United States
U.S. EPA	United States Environmental Protection Agency
UV	ultraviolet
V	volt(s)
VOC	volatile organic compound
W	watt(s)
WTP	water treatment plant

SYMBOL	**DEFINITION**
°C	degree(s) Celsius
°F	degree(s) Fahrenheit
%	percent

IWA Publishing's authorised EU representative for General Product Safety Regulations is Diane D'Arras, 15 rue Duret, 75116 Paris, France, e-mail: safety@iwap.co.uk.

Printed and bound by CPI Group (UK) Ltd, Croydon, CR0 4YY
17/09/2026
02218816-0001